Questions Raised By Quolls

Harry Saddler

16pt

Read How You Want
LARGE PRINT BOOKS, BRAILLE & DAISY

Copyright Page from the Original Book

Affirm press
books that leave an impression

Published by Affirm Press in 2021
28 Thistlethwaite Street, South Melbourne,
Boon Wurrung Country, VIC 3205
affirmpress.com.au
10 9 8 7 6 5 4 3 2 1

Title: Questions Raised By Quolls / Harry Saddler, author.

NATIONAL LIBRARY OF AUSTRALIA
A catalogue record for this book is available from the National Library of Australia

Cover design by Christa Moffitt, Christabella Designs
Typeset in Granjon by J&M Typesetting
Printed and bound in China by C&C Offset Printing Co., Ltd.

Dedicated to my parents, Hugh and Marilyn – thank you for raising me the way you did.

And in fond memory of Loretta Hemensley (1945–2021) who, along with her husband Kris, was one of the first people to welcome me into Melbourne's literary community.

Author Note

Thank you to everyone who so generously and readily shared their knowledge and expertise to help me write this book. In no particular order: Nicholas Dexter, Dion Maple, Shane Sturgeon, Gavin McLeod, Tyson Simpson-Brown, Mark Sawa, Rob Brewster, Guy Ballard, David Hamilton, Chris Jolly, Ella Kelly, Belinda Wilson, and Katherine Tuft. I hope I've done all your work justice. If I've left anyone out I apologise, but know that when I suddenly remember you three days after this book goes to print I'll kick myself – and know also that your help has been appreciated more deeply than I'm able to express.

Thank you to everyone at or associated with Affirm Press but in particular Coco McGrath for her expert guidance in helping to shape this book, Lesley Halm for her diligent and thorough copy-editing, and Martin Hughes for trusting me to create something coherent out of what was a fairly wild and woolly initial pitch. Thank you to Christa Moffitt for designing a cover that took my breath away when I first saw it; it has grown richer and more complex each time I've thought about it.

And lastly, thank you to my friends, family, and colleagues everywhere who got in touch during the darkest days of the pandemic, sent something in the mail, called on the phone, or organised communal spaces online when nothing else was possible. I'm sorry if I haven't always been so good at returning the kindness. None of us are meant to get through this life alone.

1

They came at dusk, drawn to our camp site by the smell of food. We were cleaning up after dinner in the gathering dark, the pots and pans scraped empty but with traces of our meal still lingering, when we heard the noise of them: yapping calls, as if in excitement.

We were at Solomons Jewels, on the outskirts of the Walls of Jerusalem National Park in Tasmania, my father and I. I must have been in my late teens, and my father had become passionate about hiking in the island state's remote mountains and forests. As a young man, before he met my mother, before they had a family, he'd worked for a time on the roads in south-west Tasmania. Now, decades later, he'd returned to the island state, his passion perhaps reignited by the fact that his two children, my brother and I, were old enough to join him on deep hikes into the forest and mountains as something approaching equals.

First he took my older brother to Frenchmans Cap, hiking through kilometres of mud – a trip which would become notorious within the folklore of our family. Then he took me along the Overland Track, then to Mount Anne: one of my favourite photos is of us both in our hiking gear – shorts and boots and gaiters and floppy hats and old khaki shirts; my bush aesthetic then, as now, closely modelled on his – unpacking lunch on the shoulder of Mount Anne with the dizzying heights and depths of Tasmania's glacial landscape behind us. In the valley below us is Lake Pedder, which, in 1972, was infamously flooded to become a hydroelectric reservoir despite a passionate campaign to save it. In February every year or so, we'd travel from Canberra down to Tasmania for a week or more. Once on the island we'd drive or catch a bus – on one occasion we even took a light plane – to the start of our walk, and away we'd go.

Twice we went to the Walls of Jerusalem National Park. Tucked off to the side of the more famous Cradle Mountain–Lake St Clair National Park, through which the Overland Track runs, the Walls of Jerusalem is a small park – a natural amphitheatre surrounded by the distinctive columnar cliffs of Tasmania's mountains. You hike in and then you hike out again, not staying long, entranced by the landscape which, even by Tasmanian standards, is stunningly beautiful. On our second visit to the Walls of Jerusalem, hiking through pine forest and alpine meadow in February, in the snow (not an unusual occurrence in that part of Tasmania at that time of year, but still disconcerting to a mainlander) I remember feeling as though I'd somehow stepped through a portal and ended up in Europe – until a Bennett's wallaby hopped across our path. On our last night we camped at Solomons Jewels, a collection of alpine tarns scattered throughout the bush. We had pitched our tents on the rocky ground – and it was on that night that we were ambushed by the eastern quolls, those yapping creatures who came sniffing for our dinner.

Predatory marsupials, nocturnal and endemic to Australia – somewhat like cats but with long pointed snouts – eastern quolls are either a sandy colour, like the colour of the night sky above a large city, or else they're midnight black, and covering their bodies from neck to haunches are large white spots, like spotlights searching the sky. I have a photo, somewhere – from the days when you'd take a photo and hope for the best not knowing how it'd turn out till you got the film developed – of one of the quolls roaming through our camp site, agile and light on its feet: the bright circles of its eyes in the camera's flash echo the white spots on its flanks. The whole photo is almost washed out by the silver side of our tent reflecting the flash back at the camera – a photo taken hastily, such was my excitement at this nocturnal invasion. It was the first time I'd seen eastern quolls, and it's still the only time I've ever seen them in the wild.

Another trip to Tasmania, another hike, another camp site: the first night on the South Coast Track, which we walked from west to east, and for which we'd had to catch that light plane from Hobart to the starting point at Melaleuca, far up a coastal valley. As we hiked down the valley to the coast, critically endangered orange-bellied parrots flitted overhead from time to time – Melaleuca is the only place in the world where they breed. With us was Dad's old friend Geoff. It's good to hike in a group: hiking alone for a long distance in a remote landscape with unpredictable weather is dangerous. Hiking in a group also creates camaraderie, and staves off boredom: it's easy, during long, tiring hours of trudging through the bush, even in spectacularly beautiful country, to become lost in the monotony of it – another footstep, another raindrop. Often I'd have a song, or a fraction of a song, stuck in my head for days – I'd come to wonder at the start of each hike what that trip's song would be – so having other people to point out something interesting, or something beautiful, or just to chat to, can be invaluable.

On our first night camping at Point Eric, halfway along Cox Bight, the South Coast Track's first taste of sea, a spotted-tailed quoll – the largest of the six species of quoll (four in Australia and two in Papua New Guinea) – suddenly appeared from out of the coastal heath. It stole a fruitcake that Geoff had made, taking it back into the dense vegetation behind the camp site, aluminium foil wrapping and all, before any of us could do anything about it. It had ventured right into Geoff's tent, following its nose. We'd each had only one slice of the cake, and it was delicious – and so we spent all the next week on the hike thinking about it.

Quolls were the first marsupial carnivores I ever became aware of, as a child fascinated by animals of all kinds. My father had a copy of the Australian Museum's *Complete Book of Australian Mammals,* full of glossy photos and lengthy descriptions, and I'd pore over it: to this day I vividly recall a photo of a quoll feeding on a dead parrot. Looking at the photo again now as an adult, it's still a striking image: the brown and white-spotted fur of the quoll contrasting with the bright green and deep red of the parrot's feathers and blood. And I can still recall the effect it had on me as a child: though it seems obvious in hindsight, I hadn't realised that Australia had carnivorous mammals. Carnivores were the charismatic stars of many of the wildlife documentaries I absorbed when I was growing up – lions, cheetahs, bears, and all the other meat-eating megafauna of Africa and the Americas – but until I saw that photo of the quoll, Australia's mammals had seemed timid by comparison: all kangaroos and wallabies, koalas sleeping in trees, and wombats grazing on grass, fleeing at the first sign of danger. In hindsight, seeing that photo of the quoll – a western quoll, as it turns out – started to change how I saw the mammal fauna of this continent, its richness and diversity.

At the same time, my parents, and particularly my father, were fostering in my brother and me a love of nature and the outdoors, slowly, patiently. As a child I'd kick and scream and cry whenever they dragged me out of the house to go bushwalking in the forests around Canberra, and then as soon as we arrived where we were going and had set off on our day-walk, my brother and I, playing at being bushrangers, would race ahead and hide behind a tree around the next bend, leaping out and yelling 'Ambush!' as our parents passed – both of them pretending to be taken by surprise despite how many times they'd been through this exact same thing. Not quite slow and placid enjoyment of the bush, then – but over the years, my parents' and particularly my father's quiet insistence on slowing down, breathing in deep the smells of the forest, on noticing the fine details of the bush, and listening to its sounds, rubbed off on me.

One place, close to home, where Dad and I would go walking sometimes was Mulligans Flat. A nature reserve on the northern edge of Canberra, it's a huge patch of remnant yellow-box woodland, and before Europeans arrived eastern quolls roamed there wild – as they did across the entire southeastern corner of Australia, from south-east Queensland into New South Wales and south down through Victoria into Tasmania, and then west to the edge of South Australia, until they were declared extinct on mainland Australia in the middle of the 20th century. The ecological importance of Mulligans Flat has been recognised since at least 1992, when community groups first proposed the conservation of the area as a nature reserve – a proposal which was formally implemented by the Australian Capital Territory government two years later. In 2001 I spent several days camping there as part of a fieldwork unit during my undergraduate biology degree at the Australian National University (ANU), a degree I enrolled in with the hope that it might lead to a lifetime of studying animals.

We camped at the old shearing shed in the north-west of the reserve and spent our days and nights learning how to survey for wildlife: vegetation quadrants, bat detection, setting Elliott traps for small terrestrial mammals, spotlighting for sugar gliders. We got up at dawn every morning to survey for birds and went to bed in our tents late every night. Even today I remember it as the best sleep I ever had, exhausted from working all day and made drowsy by the constant fresh air and sunshine. During that field trip the fox-and-cat-proof fence – which today symbolises so much of the ongoing work at Mulligans Flat – was still several years away from being built, and the reserve was a very different place from what it is now. When I've walked there with my father, in the years since that fieldwork unit, we've always seen plenty of wildlife: an echidna right up on the northern edge of the reserve, at the boundary between the ACT and NSW; a shingleback lizard, of the uniformly dark colour morph for which the Canberra region is known; an eastern brown snake basking in the hot sunshine in the middle of one of the gravel roads through the reserve. That last one reared up and bared its fangs at us and stayed in place in the middle of the track until we were over the ridge and out of sight – the only time I've ever been, or felt, threatened by a snake.

I was in my twenties then, and, like so many people at that age, when I was away from my family I was concerned overwhelmingly with how I appeared to other people: I wanted to be appealing, and to be accepted, and so I did things that I thought might make me appealing and accepted, leaving behind things that were often marked as socially awkward or weird. There's nothing remarkable about this. As I got older I realised that what I wanted much more was to be happy in myself, and so I returned to the things that I genuinely loved: birdwatching; obsessing about nature and wildlife in general; bushwalking; camping.

None of these things were cool when I was younger and I'm not really bothered about whether they're cool now. But in recent years I've come to wonder, again not uniquely and not originally, the extent to which *who* I am – which, as in any other person, is a messy sum of what I love and value and enjoy – comes from my own will, and how much of it was instilled in me by my parents. I laugh now at the memory of my young self's tantrums whenever my parents made me join them on a bushwalk: now, I can't imagine anything better than leaving the house to go bushwalking all day long. If I'd been raised to love airshows, or sailing, or building things, or any of the other countless pursuits by which people mark their days, then those would be my passions instead – and yet the person who I would have become wouldn't have been any less himself because of it.

And did I imagine, when I was a boy bushwalking with my parents, or a young man hiking with my father, that one day I might bring my own children hiking with me through that same landscape? I can't remember. But I know that, for as long as I've been old enough to think far into the future, I've wanted to one day have a family. Now I can all-too-easily imagine myself – in my early forties as I write this, and my father the same age that my grandparents were when I was a boy – pointing out the same things to some hypothetical child that my father pointed out to me. That child has been hypothetical my whole life – now, I'm wondering: can you still call a child hypothetical if it will never exist?

***

The first person in my family to arrive on the Australian continent was an ancestor of my father's – my great-great-great-great-grandfather, a man named Michael Farrell. Born in 1797 in County Wexford in Ireland to a Roman Catholic family, he – at '5 foot 6 inches in height, of dark ruddy complexion with dark brown hair and hazel eyes' – sounds a lot like me.

In 1816 in Dublin, Farrell was sentenced to seven years' transportation. Some sources state that Farrell was convicted of stealing plates for the purposes of counterfeiting, but my father proudly declares the crime as 'sedition!' when the subject of his ancestor comes up. After Farrell's conviction he was chained on a prison hulk in Cork Harbour until, on 14 July 1816, he, along with dozens of other convicts, was sent to New South Wales. The voyage took seven months. The ship carrying them, the *Surry* (sometimes spelled '*Surrey*' – coincidentally also the name of the English county where my mother was born and raised), arrived in Sydney on 20 December 1816. A roll call of the names of its convicts is unmistakeably Irish: *Bourke; Boylan; Brennan; Byrne; Callaghan; Conlon; Corcoran; Farrell; Maher; Mclaughlin; Ryan.*

The year after Farrell was born, the Irish Rebellion of 1798 broke out, with the aim of forcing British colonial rule out of Ireland. In Wexford the rebellion was particularly fierce – and sectarian. The very first years of Farrell's life would have been lived against a backdrop of Wexford rebels being captured and transported to Australia. In 1801 the Act of Union saw Ireland officially become part of the United Kingdom of Great Britain and Ireland, with no independent parliament of its own. Irish cereal crops were exported across the Irish Sea to Britain; in Ireland a large proportion of the population, particularly the rural poor, came to rely upon potatoes as their primary sustenance. Then potato blight arrived in the country. A million people died in the famine. Twice that many were forced to emigrate. By the time of Irish independence in 1921 the country's population was half of what it had been prior to the famine.

Michael Farrell never saw the famine, though he would have seen Irish immigrants arrive in Australia, fleeing its effects. He would also have seen further convicts like himself arrive, and may have watched the ship that bore him sail into Sydney Harbour again and again: after the completion of his sentence, Farrell stayed in Australia, and the *Surry* continued to carry convicts on a dozen voyages to Australia between 1814 and 1842. The *Surry* had a crew of thirty-five, as well as a detachment of armed men whose job was to guard the convicts. On its first voyage, in 1814, an outbreak of typhus led to the ship being quarantined upon its arrival in Sydney. By contrast, when the voyage that carried Farrell arrived, the *Sydney Gazette* reported that: '[each of the] 150 prisoners ... landed in perfect health; and all combine in one sentiment of thankful approbation of the humane and indulgent treatment they experienced on the passage from the Captain, the Surgeon, and every Officer on board'.

It took thousands of trees to build a single wooden ship such as the *Surry*. Britain sourced these trees not just from its own land but from other countries, too. The environmental impact of this industry has been much debated, but in 2012 Patrick Melby from Western Oregon University argued that over the course of the Royal Navy's long period of wooden ship building, from 1200 to 1850:

> The rapidity with which new ships were constructed and decayed or were destroyed consumed domestic oaks far more quickly than they could be replaced, stripping them from the

> landscape ... Today, the apparent inability for oaks to regenerate [in Britain] may, in part, be the result of nutrient stripping due to woodland harvesting, showing that over a century and a half later, the ecosystems of British woodlands have yet to recover from the impacts of sustained harvesting. Beyond Britain's borders, trade in masts, planking, oak, pitch, and tar demanded far more from woodland sources than were ever felt at home, and stretched the Royal Navy's reach to diverse ecosystems around the world.

Ships such as the *Surry* also carried black rats with them wherever they went. Black rats are a vector for many human diseases – including ones to which Indigenous people on the Australian continent and elsewhere had never before been exposed, and to which they had no immunity. Black rats are also regarded as one of the worst invasive species in the world, detrimentally affecting many other species including in Australia and being implicated in many extinctions.

When Michael Farrell first entered Sydney Harbour, chained in the *Surry* with no idea of what was ahead of him, he would have passed Vaucluse, where, until the 1960s, the last known colony of wild eastern quolls in New South Wales lived. Newspaper articles from before the 1960s speak of attempts by the animals – called 'native cats' in English until deep into the

20th century – to survive in the new world built around them. From 1953:

> If close to the house [trees] can also fill up your gutters with leaves, rustle against the window panes ... and enable native cats to enter between the roof and the ceiling, from which there appears to be no way of exterminating the pest says the Editor of 'Construction' who lives at Vaucluse, and has already caught five native cats in rabbit traps.

Or from 1940, this report from a Mr Taylor, secretary of 'the bird sanctuaries of Nielsen Park and Vaucluse' where quolls were said to have 'destroyed many doves, honey-suckers, kookaburras and other birds':

> Many of the native cats had been trapped. Some had escaped from the traps by biting off their own legs. So numerous were they that some had been run over by cars at night in roadways outside the parks ... Mr. Taylor added that one native cat which he discovered in the garage at the back of the trust building attacked him ferociously when cornered. He eventually killed it.

Or the 'novel discovery' by Mr Harold Donald, of Serpentine Parade in Vaucluse, in 1921:

> For the past two years he and his wife had been troubled by a mysterious tapping noise on the ceiling. Last evening the knocking began again, only louder than before. Mr. Donald decided to

> investigate, and on making his way through the manhole in the ceiling discovered the culprit – a native cat, sitting on a bed of bones. The latter accounted for the mysterious disappearance of chickens from neighbouring houses. The way the animal gained admittance to its queer abode was through a hole underneath a ventilator whence it used to steal forth nightly on its raids to the fowlyards near-by.

Michael Farrell lived to be eighty-five. Despite his sentence of transportation being only seven years he never saw Ireland again. As he grew old in Sydney – married, became a father, then a father-in-law, then a grandfather, until the day he died in 1882 – I wonder what he made of the wildlife around him. From the time of his arrival on the Australian continent until the year of his death, five species of Australian mammal became extinct, two in the 1840s and then one in each decade after that.

Farrell wouldn't have been aware of the existence, and abrupt non-existence, of these unobtrusive animals, far from where he lived – but I wonder if he shook his fist at eastern quolls as they made off with his chickens; if he was woken up at night by their vocalisations outside his house; how many he trapped and killed. I wonder, too, how often he dreamed of the Ireland he'd known as a young man: of the rabbits, introduced to Australia in 1859; hares, which were introduced in 1862, though they wouldn't reach New South Wales until the 1870s; foxes, which were finally established in the 1870s after earlier unsuccessful introduction attempts; or any of the birds that would have sung from Irish trees and fields, and later Australian landscapes, too – blackbirds, skylarks, starlings. Perhaps the animals of the Australian continent intrigued him, or perhaps they confused or frightened him, or perhaps he was merely indifferent to them. Did he ever stop to think about how the quickly expanding colony was consuming the bush around it, or did he see only the opportunity to make his mark? Did it ever give him pause to reflect that a man from a colonised country, transported across the world as a victim of colonialism, had in turn embraced the colonial project on another continent, and thrived in it, and taken advantage of all the opportunities that it presented to people like him, opportunities made available in the space created by the killing and dispossession of other people in their own homeland much as he must have seen in Ireland? Or is it my job, as his descendant, to think about it all on his behalf?

As well as eastern quolls, Farrell would have been familiar with spotted-tailed quolls – called 'tiger-cats' at the time. Like eastern quolls they make numerous appearances in colonial newspaper articles and letters, providing a map of sorts of the expanding colony: a writer in *The Manaro Mercury,* printed in Cooma, reported hearing 'night owls of every description, wombats, native bears, and the monstrous and ferocious-looking tiger cat made such discordant and uncouth sounds as would terrify one more superstitious than myself'; in 1868 a 'sporting gentleman' on his travels near Strathbogie killed 'a monster tiger cat, after a most severe encounter and exciting battle, having no weapons but stones ... From the description given to us the chase and battle must have been very amusing'. And, in 1857, William Lyall – born in Scotland, raised in Tasmania, and by then resident in Melbourne and a keen 'acclimatiser' of foreign species to Australia – suffered 'an unfortunate incident ... to mar the hopes of those who desired to see the English [actually Asian] pheasant introduced and acclimatised in Victoria...': nine pheasants imported by Lyall were eaten by 'a tiger cat, who satisfied with his dainty meal, appeared to have taken up his residence on the spot where he had committed this gross outrage'. Then, in 1850, Thomas Francis Meagher, Irish nationalist and 'distinguished exile', wrote a letter to *The Cork Examiner* telling of his life in 'a wild place' in Campbell Town, Tasmania, 'stocked with kangaroo, tiger cats, black swans, and snakes'.

I've also seen all of those species in Tasmania. The island can seem like a throwback in this way – to a time before the wildlife of the Australian continent was devastated. Mammal species that are extinct on the mainland or are severely depleted are still found in Tasmania: not just eastern and spotted-tailed quolls but eastern barred bandicoots, southern bettongs, long-nosed potoroos. Many species that were abundant on the Australian continent only two centuries ago have now almost disappeared. Some of the first words that every Australian schoolchild learns about their country are of its 'beauty rich and rare', but 'rare' is an unsettlingly ambiguous word: I've always taken it in this context to refer to the uniqueness, in a global setting, of this continent's flora and fauna – but it's hard now not to see it as also indicative of the numerous species that have been made extinct since European colonisation, or that are on their way to being made so.

Another thing that we're taught from childhood in Australia is that everything on this continent was retrograde before Europeans arrived. It constantly astonishes me that a country so belligerently proud of its achievements in the world can be so prejudicially dismissive of the fact that it's home also to the world's oldest living human cultures – people whose achievements in thriving peacefully in the often harsh environments of this continent are so casually, viciously and routinely dismissed. And even our 'rich and rare' wildlife is often celebrated with backhanded compliments. In school, not so very long ago, I was taught at every turn that Australia's marsupials were primitive, as if, when it came to mammal evolution, this continent had been left behind by the rest of the world (never mind that Australia also has – and had – a wealth of placental mammals). The attitude that the Australian continent is an innately backwards place has deep roots in the history of European colonisation of this land. In the mid-19th century 'an old bushman', Horace William Wheelwright, wrote in his memoir *Bush Wanderings of a Naturalist:*

> There is no particular wild breed of cattle, horses, or sheep, indigenous to Australia. In fact, it would appear that this immense island had been left a barren waste upon the face of the globe, until its hidden resources should be developed by the skill and perseverance of civilised man; for so genial is its climate, and so peculiar its soil, that almost any animal or plant will thrive here, no matter from what part of the world it is imported. And this very fact, now clearly proved, goes far to refute the argument that Australia is a country fitted by nature only as a residence for the lowest class of animals, the marsupial.

Wheelwright spent four years living and working as a hunter in the vicinity of Melbourne, a city that had been founded only a couple of decades before his arrival from England. In his book, Wheelwright details the wildlife of the part of Australia in which he lived – wildlife that he shot to sell at market. His observations are striking for being a snapshot of a moment when native Australian wildlife was still abundant, but was on the verge of collapsing. He notes that eastern quolls – 'the *Native Cat* or *Dasyure*' – are 'one of the commonest of all the bush animals ... They are common throughout the whole bush ... one of the most prolific animals in the bush, and I have often killed six young ones in a nest.' Wheelwright describes eastern quolls as 'very destructive to birds, especially poultry, and eggs'. It's presumably for this reason that he killed them so indiscriminately.

Even among this native abundance he notes also, admiringly, that:

> On some of the islands in Western-port Bay [sic], and along the coast, the common wild English rabbit has been turned out, and thriven well ... There is much poor hungry soil in Port Phillip, which is of little use for the plough, and less for pasture, mostly scrub and sand, but where many English esculents [edible plants] would grow, if planted wild, for the rabbits to feed on. The rabbits could always be sold in Melbourne for good prices. The native cats and hawks would be their worst enemies.

With hindsight, more chilling is a passage in which Wheelwright notes that: 'The *Domestic Cat* sometimes wanders away from a station and turns bushranger; and certainly the largest cat that I ever saw in my life was a large black and white one which I killed in a honeysuckle scrub here.'

I wonder if Wheelwright, or my ancestor Michael Farrell, noticed as the wildlife of the Australian continent diminished around them? Since European colonisation, more than a hundred Australian species of plant and animal have been made extinct. As the CSIRO's Andy Sheppard and Linda Broadhurst write in *The Conversation:*

> Australia has the highest rate of vertebrate mammal extinction in the world, and invasive species are our number one threat.
>
> Cats and foxes have driven 22 native mammals to extinction across central Australia and a new wave of decline – largely from cats – is taking place across northern Australia. Research has estimated 270 more threatened and endangered vertebrates are being affected by invasive species.

***

Michael Farrell's descendants – including my family – have spread themselves wide across the southern part of the Australian continent. My paternal grandfather was born and raised in Tasmania before moving with his parents to Adelaide, where my father and his sisters grew up. My parents met in London, before living in Sydney, where my brother was born, and then finally making a permanent home in Canberra, where, in 1979, I was born. Canberra is around 1200 kilometres from Adelaide – but with my mother's family all in England, my father's family were the closest to home, and so every Christmas my mother, father, brother and I would all pile into the car and drive to South Australia – sometimes with our family dog.

In the early years we did the trip overnight, staying in a motel somewhere near the Murray River on the border between New South Wales and Victoria. Then in later years when my brother and I were older we'd all get up at dawn and drive straight across in one day. Before we had a car with air-conditioning we tried any way we could think of to keep ourselves comfortable in the Christmas-time heat: damp towels hanging from the windows; black plastic screens attached to the windows with suction caps to keep the sun at bay. We'd moan and complain by rote about the monotonous Hay Plains, and eat lunch under the straggly shade of eucalypts in the heat of a late December day while peaceful doves called *olly-ock* from unseen perches and emus wandered by in the hazy distance. We'd listen to the radio the whole way, or to tapes or, later, CDs, all the usual Christmas favourites blaring out of the tinny car speakers, including once – and I'll always remember it – when the 'Hallelujah' chorus of Handel's *Messiah* started the instant we turned into my grandmother's street. A sleep-in the day after the drive would be the most blissful thing in the world, the smell of the warm Adelaide night after fifteen hours breathing car-scented air the greatest medicine imaginable.

The Hay Plains that so bored my family and me when we drove across them every Christmas were once home to areas of native woodland and shrubs.

Over-grazing by sheep, introduced by European colonists, created 'secondary grasslands' – areas of depleted biodiversity, not regarded as natural grasslands. With the plants went animals, including the northern hairy-nosed wombat, which according to a report prepared by the New South Wales National Parks and Wildlife Service 'was once found throughout southwestern NSW, particularly in the Riverina Bioregion. However, competition with cattle, sheep and rabbits has led to the rapid decline of the species ... and the wombat is now presumed [locally] extinct'.

Flat land is easily cleared, and Australia – particularly inland Australia – has a lot of flat land. At the end of the 19th century, as sheep were eating and trampling the vegetation of the Riverina Bioregion in New South Wales, colonists were also clearing the Mallee, a vast semi-arid belt of low woodland that spans a huge swathe of southern Australia, from New South Wales through to Western Australia. The invention of the 'stump-jump plough' in 1876 facilitated the devastation of this ecosystem: where previous ploughs had broken on the dense, tough lignotuber roots of Mallee trees, the stump-jump plough did exactly what its name suggests. Once colonists got a foothold in the Mallee they went to war with it, felling trees and burning regrowth until nearly all of the Mallee – along with its unique flora and fauna – was destroyed: as much as 65 per cent of the Mallee in Victoria and 75 per cent of the Mallee in South

Australia, according to the Australian National Botanic Gardens.

When I sat in the back seat of my parents' car, driving across inland New South Wales and Victoria to South Australia, I used to stare out the window at the remnant scraps of Mallee woodland along the roadsides. I'd be hoping against hope that I might somehow catch a glimpse of a malleefowl, a legendary bird at the top of many birdwatchers' must-see lists. If I'd known about it at the time I would also have hoped, in my childhood naivete, to see a western quoll in the twilight hour – though that would have been even less likely than seeing a malleefowl. The western quoll was once found across a huge area of Australia – 70 per cent of the continent, in fact – from western Queensland and New South Wales to the coasts of South Australia and Western Australia. It was absent only from the high rainfall areas of the east and south-east and the tropical north, and 'relatively abundant' according to a fact sheet compiled in the early 2000s by what was then called the Department of the Environment and Heritage. However, by the time of my family's drives, it was found only in the southwestern corner of Western Australia – a mere 2 per cent of its original range.

The scope of this loss is written in the western quoll's other common name: 'chuditch'. The chuditch is the only one of Australia's four quoll species to be widely known by an Indigenous name – in this case, a Noongar name. The Noongar are the Indigenous people of south-west Western Australia, the chuditch's last stronghold. The website of the South West Aboriginal Land and Sea Council explains that there are fourteen different language groups within the broader Noongar cultural bloc. But at the time of European colonisation, there were *hundreds* of languages spoken on this continent, from coast to coast and throughout the interior. The animal that we now know as either the western quoll or the chuditch would once have had scores of different names. As colonists cleared vegetation, and as the animals they introduced spread throughout the land, damaging habitat and devouring wildlife, the chuditch's once-vast range shrunk and shrunk to almost nothing; at the same time, Aboriginal people and their cultures were eroded and shattered in a campaign that mainstream Australia is still unwilling to call genocide.

Now, of all those hundreds of languages across Australia, according to the Australian National University's Jane Simpson, only thirteen are still spoken by children. Innumerable names for the chuditch, the western quoll – and names, too, for its environment, its lifestyle, the ecological relationships within which it lived, the cultural history its presence spoke of – are going or gone.

This destruction of habitat, and of species, and the cultural destruction that went along with it, is a familiar story in Australia, as in other colonised countries, but we're taught at every turn to value the 'pioneering spirit' that so devastated the land and people that stood in its way – while also ignoring the resilience of Indigenous people who everywhere are still fighting to save their cultures.

When I think about the animals that have disappeared from the Australian continent and its surrounding islands – the thylacine, its last survivor pacing hopelessly back and forth in its concrete cage and left to freeze to death; the Bramble Cay melomys, slowly starved of food as climate change pushed the sea around its tiny island home higher and higher; the numerous native and unique rodents all-too-easily hunted by cats and foxes throughout Australia, etc., etc. – it feels emblematic of everything else that colonisation has destroyed, or attempted to destroy, and I feel an almost unbearable weight of loss, a lead-like heaviness in my heart. I ache for the absence of these unique species, each inherently deserving of their place on earth, and I ache for the lost or fragmented knowledge of them and of their place in the vast interconnected web of life.

If they were considered at all by Australian colonialist thinking, these species were viewed as nuisances at best, with catastrophic results. The story of cane toads is demonstrative: a sugarcane industry was established in Australia in the second half of the 19th century, but sugarcane growers struggled with beetle larvae eating the roots of the cane. By the early 20th century, sugarcane had become such a significant industry that a solution to this consumption by beetle larvae was being urgently sought. The solution arrived at was to introduce a supposed predator: the cane toad. In late 1935 in northern Queensland, 2400 cane toads were released. Only fifteen years later, having had no effect on the beetles, cane toads were declared a problem species. This is a familiar story; what's less well known is that the beetles that the cane toads were brought to Australia to destroy were *native Australian species.* The presence of these naturally occurring species was deemed not just less important than the national economy, but an active hinderance to it. As always capitalism and colonialism go hand in hand.

***

I live in Melbourne now. I've lived here since 2004. I only meant for it to be for a year or two. When I was growing up in Canberra, the question hanging over every young person in the city – sometimes spoken, sometimes unspoken – was 'where are you going to move to?' Despite being Australia's capital, Canberra is often described as a large country town, and that description is never more apt than in the assumption that its young people will go and live somewhere else once they achieve independence. For a long time that 'somewhere else' was Sydney, only three hours up the Hume Highway. My brother, four and a half years older than me, moved there for a while, as did all his friends, as did generations of Canberrans before him – to be replaced by young people coming the other way, attracted by graduate jobs in the public service. An ongoing cycle. By contrast, Melbourne – twice as far and two states away – may as well have been the Emerald City from *The Wizard of Oz.* Growing up I wasn't aware of meeting anybody who'd ever been to Melbourne; I'd never even knowingly seen photos of it, let alone had any conception of it as a place other than understanding that they played 'Aussie Rules' there. But then, suddenly, in the early 2000s, something happened: Melbourne became *the place to be.*

I was told later that the phenomenon is called 'chain migration': all my friends moved to Melbourne, so I followed them. And though Canberra, with its mountains and bush and clean, crisp air still felt like home – as it still does today – after returning from a year living in England with my maternal grandparents after I graduated from uni, taking frequent train trips into London only 45 minutes away from their Surrey house, my home town suddenly felt small in a way it hadn't when I was at school, or studying at the ANU.

I broke the news to my parents, and to my girlfriend of the time, and I booked a ticket on the Canberra-to-Melbourne train (because I'd be able to take more luggage that way), packed two big suitcases and off I went. Though I didn't know it at the time, I was travelling from Ngunnawal country, through Wiradjuri, Waveroo and Taungurong country, to Woiworung country. One of my Adelaide cousins from my dad's side of the family was living in North Fitzroy at the time, so I lived with him while I looked for a place of my own.

I was living near Merri Creek, one of the main tributaries of the Yarra River – as I still do today. Not far away was Yarra Bend Park, Melbourne's largest bush park. Rakali and platypuses still swim in Melbourne's waterways; kangaroos are abundant in the bush of the outer suburbs and sometimes even come as far downriver as Yarra Bend Park. Once recently while walking through the park I saw an ant's nest in the grass, ripped apart in a way that made me certain that an echidna had passed that way. Possums are abundant and there are even, in some places, sugar gliders: at the end of 2018 I regularly caught a bus up the Eastern Freeway to Westerfolds Park, on the banks of the Yarra in the suburb of Templestowe, tracking and watching a family of gliders. Eventually I found the hollow they lived in and could watch them at dusk, as they leaped out into the surrounding trees one after the other, sometimes right over my head. If you look closely in the parks up and down the banks of the Yarra you can often find wombat burrows – and if you're out after dark and you're very lucky you might even see the wombats that live in them.

Quolls, too, once lived here. Before I even moved to Melbourne, on one of my first visits to the city, friends took me to Studley Park. Not knowing the city at the time, I had no conception of where it was, but it felt like miles from anywhere: surrounded by bush, next to the wide Yarra River. In fact it's in the inner city, just over the river from Collingwood, where one of Melbourne's oldest football teams was founded, and only a few kilometres from the CBD. Studley Park is famous for its boathouse, built in 1863 and in continuous use ever since. Nowadays it's a popular spot for picnics, and on the opposite bank from the boathouse is an oval and a golf course. A small area adjacent to the boathouse has been cleared for lawns, but old eucalypts still stud the grass, and beyond the lawns – particularly on the steep escarpment behind the boathouse – bush still dominates.

From Yarra Boulevard at the top of the escarpment, various dirt tracks, both official and unofficial, spider-web down into the bush and back to the river. Along the water's edge huge river red gums form a canopy; below them is an understorey of smaller trees such as casuarinas and wattles, and below those a layer of shrubs and bushes: sweet bursaria, hop goodenia, tree violet. Native small-leaved clematis covers the ground and climbs up everything. River red gums are infamous for dropping branches, and the ground is scattered with woody debris: everything from sticks to logs, many of them hollowed out by termites or rot. If you can tune out the traffic noise of the nearby Studley Park Road and Eastern Freeway you can easily forget that you're in the inner suburbs of Australia's fastest-growing city.

Studley Park is also where one of the last known wild mainland eastern quolls in Victoria survived. An article in Melbourne's *Argus* newspaper from 1956 states that: 'there is an isolated colony in one of the wilder parts of Studley Park; and every now and again the body of one is brought to the [National] Museum after being dazzled and knocked over by a car at night on Studley Park Road, or the Yarra Boulevard'. How long that colony persisted for is difficult to determine, but it's believed that by the late 1950s the species was extinct in Victoria.

Because extinction is by its nature so final, it's easy to imagine that it must be obvious: the loss of an entire species is so enormous a cataclysm that it must surely reverberate throughout the world. But many species slip away quietly, their existence mostly unremarked upon and their disappearance unnoticed until decades later.

Even for well-known and charismatic species such as the eastern quoll – once 'one of the commonest of all the bush animals' in Victoria, according to colonial accounts relayed by the website *eMelbourne,* the online version of *The Encyclopedia of Melbourne* – the very vastness of the Australian continent holds within it the possibility that the animal might still exist somewhere, even if one hasn't been officially seen in decades. In 2019 a review of reported eastern quoll sightings on the Australian mainland in the years since they were declared extinct everywhere except Tasmania listed claims of animals seen throughout NSW: on several occasions on the Central Coast in the 1990s; in Wollemi National Park, in the Blue

Mountains west of Sydney, in 2002 and 2006; and in the far south of NSW in 2013. The paper concludes that: 'There has been insufficient recent mammal survey effort to definitively support these public reports but at this stage there are sufficient recent credible records to consider that this species may not be extinct on mainland Australia.' You don't know what you've got till it's gone, as the saying goes – to which we could add: sometimes you don't even know *if* it's gone.

Walking around Studley Park now, with the knowledge of what was once there, I feel the absence – and once I notice that absence, I start noticing other absences, too. For all that the bush in some parts of Studley Park appears pristine, it seems to me to be a surprisingly empty bush; other reserves and bush fringes in inner Melbourne bustle with the chattering of small birds: groups of brown thornbills working their way through the lower branches of trees; families of superb fairy-wrens and white-browed scrub-wrens calling to each other as they forage in low bushes and on the ground for any invertebrate they can find; flocks of silvereyes flying from tree to tree, contact-calls pinging constantly back and forth between the birds; tiny pardalotes *pip-pipping* from the very tops of the tallest eucalypts, where they're chased by white-plumed honeyeaters. Yet more often than not when I walk through the bush at Studley Park, even at dawn when bush birds are at their most active, the trees and shrubs are conspicuously silent. Only larger birds, those that are robust enough in size or temperament or social structure to not just survive but thrive in the inner city, can be relied upon: rainbow lorikeets, noisy miners, occasionally grey butcherbirds. Every now and then I'll stumble upon a group of the smaller birds, and sometimes even an eastern yellow robin, but for the most part although Studley Park is to my eyes one of the most beautiful patches of bush in Melbourne it seems to me also like a ghost habitat – or perhaps more accurately, a regenerating one: hidden in the undergrowth next to one of the many tracks that crisscross through the park is a wooden sign saying 'FRIENDS OF THE YARRA Re-vegetation Project 1986–'.

I can imagine quolls hunting at night in Studley Park, and returning at dawn to shelter through the day in the hollow logs on the ground, the fallen river red gum branches. But when I look more closely at those logs I notice that they're overgrown with oxalis, a choking weed that covers the ground at Studley Park like a carpet. When I think about the oxalis – when I worry about how it may have got there, spread as a seed on the tyre of a mountain biker or the shoe of a walker such as myself – and about how it might be eradicated from such a wild yet popular area, the whole history of environmental destruction in Australia since colonisation, the entire dizzying struggle of the manifold environmental catastrophes now raging across the planet, seem encapsulated in this one tiny plant, in this one tiny part of the world. And then when I extrapolate out from that, when I try to imagine the entire continent or even the entire planet, what young people growing up now or those growing up in the coming decades will have to face, the vastness of the crisis that we're now in becomes so dizzying that my mind grinds to a halt until I'm left with just one paralysing thought: how do we go on?

2

My dad still goes hiking today, sometimes. Mum likes going bushwalking, too, but overnight hikes and camping have never been her thing: that's always been more Dad's activity of choice, and by his example mine, too. There's a kind of deep, slow immersion in place that you only get through walking: one step at a time, one kilometre at a time, one day at a time, plodding gradually into a landscape where cars and trucks and bikes and planes can't go. Now that I live here in Melbourne, I've found places where I can go for long day-walks. There's a bush track along the Yarra that follows the river's winding course from Fairfield to Abbotsford, sometimes high above the river and sometimes only centimetres from the water's edge. If you drove along Yarra Boulevard, which follows a similar route, it would take you less than a quarter of an hour, but walking the full length of the track takes about four hours. Further out, with the recent extension of the South Morang suburban train line to Mernda, it's easier than ever to walk through Plenty Gorge, and on the last day before Melbourne entered stage-four lockdown against the coronavirus pandemic in July 2020 (when exercise became limited to only one hour a day within a 5-kilometre radius of home) that's exactly what I did: my housemate had recently tested negative to the virus, and I felt fine, and I knew from previous experience how empty of people Plenty Gorge could be, so I went for a four-hour hike. Wallabies, kangaroos, wombats; birds aplenty. I'd been hoping to replicate, sometime in 2020, a hike I'd done in 2019, walking the length of Plenty Gorge Park, connecting up with the Plenty River bike trail and then walking the length of that, too, to the Yarra: 30 kilometres all up, which took me all day. While 2020 had other plans, Plenty Gorge remains one of my favourite places in Melbourne to go walking.

In early September 2020, it was a week shy of my 41st birthday. In a normal year I'd have been travelling back to Canberra to see my family, but that wasn't possible now. Earlier that year, in May or June, my parents had spoken, hopefully, of hiring a campervan and touring Victoria in July, and coming to see me – but lockdown and border closures and Victoria's second wave of infections quickly put paid to that idea. By Christmas 2020 when the borders briefly reopened and I was finally able to travel interstate for a week it had been nine months since I'd last seen my parents and my brother, though we'd often spoken on the phone – their lives in Canberra gradually opening back up like flowers in the sun, mine in Melbourne stuck in place as if pressed between pages. But my situation was far from unique.

The previous Christmas, 2019, instead of me travelling to Canberra my family all drove down to spend the holiday in Melbourne. It was a plan put in place months in advance, but by the time they did it the chance to get out of Canberra, and away from the yellow bushfire smoke that had smothered it for weeks on end, was a relief. Melbourne, by contrast and good fortune, experienced very few smoky days in the long fire season of 2019–20, but one of those days was when my family arrived, after an eight-hour drive. 'I was looking forward to getting away from the smoke,' Mum told me wearily when they arrived.

For her Christmas present I took Mum to the Basquiat/Haring exhibition at the National Gallery of Victoria; and for Dad's present, I took him walking in Plenty Gorge Park. Just a short walk, relatively speaking, from Hawkstowe station to Middle Gorge station – only a couple of hours – but representative of what makes the gorge so special: the meandering river; the rock faces; the eucalypts and wattles and complex understorey; the birdsong; the impossibility of believing that these miles and miles of bush are in the outer suburbs of a city of five million people, only three quarters of an hour by train from the CBD. His friend Geoff now lives in Melbourne, too – has done for years – and so we invited him to join us. Neither he nor Dad had ever been to the gorge, and though I'd tried to explain the route of the trail to Geoff in an email, the gorge can be confusing – there's one spot in particular where a creek flows into the Plenty River where I reliably make a wrong turn every single time – so it made more sense for him to join us, and see the way for himself. Plus of course it was nice for my dad to spend a few hours with his old friend, walking together through the bush like they always used to.

On our hikes in Tasmania it was Dad who always led the way – the organiser, the map-carrier – and on that day in Plenty Gorge it didn't escape my attention that the roles had now reversed. This is the nature of getting older, and of parents and children, as we all know: the roles reverse; the carer becomes the cared for. At the start of 2019, on a Sunday morning, I was woken by a phone call from my mum: she'd been due to travel from Canberra up to Sydney where she was scheduled to have a knee replacement, surgery which would be followed by days of recuperation in the hospital. The plan had been for Dad to go with her, and keep her company through that recuperation – but he'd somehow, dramatically, injured his eye during the night (at the hospital he was told it looked as though somebody had kicked a football into it) and had to have his own emergency surgery. So while my brother looked after Dad in Canberra, I flew that afternoon from Melbourne to Sydney to be with Mum. It happens gradually, but also suddenly.

I have a photo of that short walk through Plenty Gorge with Dad and Geoff at Christmas time in 2019. Down the slope from Nioka Bush Camp, which is sometimes used by school groups, there's a wooden bench near a bend in the river where tall manna gums and candlebarks grow side by side. The photo is of my dad and Geoff sitting next to each other on the bench, gazing at the bush, talking – I don't know what about; it doesn't matter. Two old friends, each decades older now than when they first met, their children grown up; living in different cities but still connecting when they can. The bush is dry and yellow-looking, as bush often is in late December. On the phone earlier that month Dad had talked about how some of the places that burned in the fires that season – in some cases entire national parks – would never recover in his lifetime, and I recalled silently a walk we'd done years earlier in Namadgi National Park, to Yankee Hat, where what's often described as the ACT's only known Aboriginal rock art is located: it had been preparation for one of our Tasmanian hikes, and our plan for Yankee Hat had been to hike overnight, but regrowth from a fire fifteen years earlier had been so dense on the mountain that it was impenetrable and we'd been forced to turn back. Walking through boggy grassland at dusk as the kangaroos emerged and sleet began to fall, we returned, defeated, to Dad's car. Another time, also in Namadgi, we'd again hit regrowth that had slowed our progress to such an extent that we eventually had to camp in scrub and dig through cutting-grass where the map said a creek bed was to be found until we hit groundwater. By the time we finished setting up the tent it was already dark, and we were too tired to cook, so we went to bed with only a couple of slices of fruitcake each for dinner.

Only a few days after our Christmas 2019 walk in Plenty Gorge, a fire broke out at the southern end of the gorge. It was kilometres from where we'd been, but I watched the online updates anxiously. Eventually it was contained, but the cleanup work took many months. At the start of 2020, before the pandemic, I'd wanted to take some friends swimming in the Blue Lake at the southern end of the gorge – an old quarry dug right next to the river, which unsurprisingly flooded – and had noticed, late, that the lake was in the closed part of the park. I'd rung Parks Victoria to confirm. My friends and I were desperate for a wild swim, a walk in the bush – anything that felt normal after so many months lacking normality and at a time of year when people in southern Australia spend every spare hour they can outside. The Parks Victoria employee who answered my call explained that it was necessary to assess each tree damaged by the fire for its stability and safety before it would be possible to reopen the southern part of the gorge. On 26 June 2020 Parks Victoria released an update on its website, which explained that:

> Wurundjeri people have, on a seasonal basis, visited the area now known as Plenty Gorge Park for tens of thousands of years. The fire and its suppression work have exposed important, and previously unregistered cultural material that is associated with this occupation.

> Parks Victoria has engaged a Cultural Heritage Advisor with assistance from Wurundjeri to survey the affected area. Unregistered cultural material located during the survey is now being formally registered.
>
> Parks Victoria continues to work with Wurundjeri to obtain the necessary permits to rehabilitate the six kilometres of bulldozer-built containment line that was constructed during the fire suppression effort.
>
> Although it may not appear that much is going on from afar, the detailed survey and recording process is taking place, preserving history for future generations...

In January 2020 it was reported that fires in Budj Bim Cultural Landscape, 300 kilometres west of Melbourne, had also revealed previously hidden elements of an ancient system of stone channels built by Gunditjmara people thousands of years ago to catch and harvest eels. The Budj Bim aquaculture system – what was then known of it – had been added to the UNESCO World Heritage list in only July 2019.

In Plenty Gorge there's a scar tree – the only one that I'm familiar with – just up the slope from where I took the photo of Geoff and my father and, though I make sure to visit it whenever I'm walking through the gorge, I've never photographed it, or touched it (I'm a compulsive toucher of trees). To do so would feel inappropriately possessive: my people have already taken so much from Indigenous cultures on this continent, erased so much of their presence, and it feels wrong to me to own even a memento of something so ancient, something that belongs to a culture not my own. It's enough for me that I know where it is, and I always show other people when I'm taking them there because I want everyone to remember that every part of the Australian continent has a human history older than most of us can conceive of.

When I walk through Plenty Gorge alone, and see only a handful of other people, or even when I go there as I did before lockdown explicitly because I *know* I'll see few if any other people, I have to remind myself at the same time what that emptiness means. Hikes in many parts of Australia are often described as hikes in the wilderness, and the term 'wilderness' is even incorporated into the placenames of many areas of the Australian bush, but on this continent to speak of *wilderness* – 'land which is wild, uncultivated, and inhabited only by wild animals; a wild, uncultivated, or uninhabited region' – is to erase the millennia-old and ongoing presence of Indigenous people and their nations and cultures. Ideas can be as invasive, and as destructive, as foxes or cats. It's comparatively easy to remove the notion of 'terra nullius' from legal doctrine; removing it from the mainstream national culture is infinitely harder.

That new emptiness has an effect on the bush. As anyone who has even the most basic knowledge of the traditional fire methods of this continent's Indigenous peoples knows, fire in Australia can be a regenerative as well as a destructive force. But the fires we have now aren't what this continent's organisms evolved to survive. The regenerating bush that my father and I got stuck in in Namadgi was a good thing, as much as it obstructed our passage and forced us to change our plans – but fires now don't happen every fifteen years. They happen much more frequently than that; a frequency with which the Australian bush is not equipped to cope.

More worryingly still, fires now happen in places that have never burned before. In August 2018, on a trip to Brisbane, my friend Andrew took me birdwatching in the rainforest of Lamington National Park, on the Queensland–NSW border, but the birds were absent, pushed further up into the mountains by the extreme dryness in the rainforest. Exactly a year later, in August 2019, I was in Brisbane again and once again Andrew took me birdwatching, this time into the Lockyer Valley to look for waterbirds – but there was no water. 'It's as dry as a chip,' Andrew said, again and again.

In September 2019, only a month after that trip to the Lockyer Valley, a fire broke out in Lamington

National Park. Binna Burra Lodge, only 6.5 kilometres north-east of where Andrew had taken me, was burned to the ground. This happened ten months after rainforest had burned in central Queensland, in November 2018 – an event that was described by journalist Mark Doman as 'unlike any fire Queensland had faced in recent memory'. With typical Australian understatement, Associate Professor Rod Fensham from the University of Queensland was quoted by the ABC in an article about the September 2019 fires as saying, 'You have to wonder about the effects of global climate change on our climate.'

By Christmas time in Plenty Gorge, Dad, Geoff and I – and everyone else in Australia – had all been watching wearily and with increasing anger and despair as fires raged across millions of square kilometres, sometimes in areas that had never burned before, sometimes torching the same areas over and over again. At the same time, disinformation and conspiracy theories burned almost as quickly online – the trademark conflagration of the early 21st century. So Plenty Gorge, on that short day, was something of a respite, and it made me happy, if somehow bittersweetly so, to see my dad and his old friend sit on the bench together and look at the trees that neither of them had seen before.

After a while I went and sat with them, and Geoff brought out his homemade fruitcake, just like the one that had been stolen by a spotted-tailed quoll on the first night of our walk along the South Coast Track twenty years earlier. I've often wondered about that quoll since: if the carnivorous marsupial found the sustenance it was looking for in the fruitcake, and whether it tore off the aluminium foil that wrapped the cake or whether it just ate the whole thing, foil and all; whether the foil passed straight through its gut undigested or whether the quoll suffered as a result. 'I still remember that bloody quoll,' Geoff said as we now ate our slices of cake in Plenty Gorge.

Aluminium foil or not, that 'bloody quoll' would be long dead by now: if it went on to have offspring, and those offspring had offspring, etc., we'll be half a dozen generations deeper into that quoll's ancestral line. Needless to say, fruitcake is outside a spotted-tailed quoll's usual dietary preferences: 'birds, reptiles and mammals such as bandicoots, possums, echidnas and rabbits', according to the former Department of the Environment and Heritage's quoll fact sheet. In huts on the popular Overland Track, when Dad and I walked it, the visitor books were full of anecdotes about notorious possums – one book had entry after entry, from visitor after visitor, recounting the exploits of 'Black Pete' – but if our particular quoll was a regular presence at that camp site on the South Coast Track, there was no record of it. 'They can be locally abundant but quite patchy in their distribution,' David Hamilton of the University of Tasmania tells me in an email. Hamilton studies quolls and their larger relatives the Tasmanian devils. In April 2020 he made a music video using felt models of the various quoll species to teach people how to differentiate between them, and he's a lively and amusing user of Twitter, where he has a fondness for quoll puns – you can find the music video on *YouTube* under the name 'Status Quoll'. And endearingly, he calls spotted-tailed quolls 'spotties':

> [W]et forest is their preference, so there are high numbers in the [north-west of Tasmania]. We see spotties in some numbers at lots of other [Tasmanian devil] monitoring sites though, so they occur in Freycinet, wukalina (NE) and the Midlands, where it's much drier, as well as places like Cradle [Mountain].

I'm used to thinking of Tasmania as a safe-haven for quolls and for other native animals, although every visitor to Tasmania is familiar with how much roadkill is on the roads there – 300,000 animals are killed every year, or 32 every hour – and it's an open question as to whether this roadkill indicates a particularly high number of animals in the Tasmanian bush, or a particularly cavalier attitude among Tasmanian drivers, or both. I ask Hamilton how many spotted-tailed quolls there are in Tasmania and he tells me, 'I can't say I've heard of an attempt to make a systematic population estimate of [spotted-tailed quolls] in Tassie recently, so I'm not sure on that one,' but he adds that, 'I think their numbers are quite stable – we haven't had any notable declines in recent years, and they seem to do OK in more disturbed areas like agricultural fringes or forestry plantations.'

Predators such as quolls, or Tasmanian devils, can be vulnerable not just to being hit by cars – as are other animals – but also to becoming 'secondary roadkill': attracted onto roads by the free meals animal carcasses already killed by car-strike offer. It's another startling way in which the modern world we've built is incompatible with the way so many animals live. As Ann Jones put it on Radio National's *Off Track,* 'The carnivores get busy chowing down, their heads buried in a rib cage and they don't hear the cars coming. Until it's too late.'

I ask Hamilton about the population of eastern quolls in Tasmania – which really is their stronghold, being the only place they've never been declared extinct. He tells me that they can be locally abundant but are patchily distributed in the state, and are absent entirely from the west of Tasmania and on the north coast beyond Burnie. As for their overall population:

> I believe the last population estimate for Tassie was somewhere in the order of 10–12,000, but that's not overly robust. They had a steep population decline in the noughties, completely disappearing from some areas and with large drops in others ... In the last 3 years or so however, we've seen large upticks in the number of Eastern quolls caught at some of our monitoring sites (namely Cradle Mountain & the Channel

region), so they appear to be making a recovery in some areas at least.

Aussie Ark, an organisation on the Central Coast of NSW – originally created in 2011 with the intention of saving Tasmanian devils and now dedicated to conserving numerous threatened Australian species – states that: 'Across Tasmania, [eastern] quoll numbers have declined by more than 50 per cent in the 10 years to 2009.'

Tasmania's status as a last haven for wildlife that is extinct or endangered on the mainland is often put down to the fact that foxes have never become established there. This doesn't mean, though, that native animals have never been persecuted on the island. Clearing of forest is a major threat to Tasmanian wildlife to this day, as is the ongoing clearing of habitat on the mainland. And also, as on the mainland, native animals including quolls were hunted in earlier colonial times. There are so many ways for an animal to go extinct. On 22 February 1900 my ancestor Michael Farrell's great-granddaughter (my great-grandmother) Lilian Allder married Walter Saddler in Melbourne; newspaper reports of the wedding noted the couple's intention to move to Hobart. In the following decade three species of Australian rodent would go extinct in quick succession: the long-tailed hopping mouse on mainland Australia, the last specimen of which was collected in 1901; followed by two species of rat on distant Christmas

Island – the bulldog rat and Maclear's rat, in 1903 and 1904.

Between Michael Farrell's arrival in Sydney in 1816 and Lilian Allder's marriage in Melbourne in 1900 seven species of Australian mammal, found nowhere else in the world, had been driven to extinction, including such diverse and unique species as the white-footed rabbit-rat, the broad-faced potoroo, the eastern hare-wallaby, and the short-tailed hopping mouse. In April 1878, only a couple of decades before newlyweds Lilian and Walter moved to Tasmania, *The Tasmanian* newspaper had reported on the arrival and selling of various kinds of cargo in Hobart, including hides: 'native cat' skins were priced at two to three shillings per dozen. (The newspaper also quoted the going rates for the skins of kangaroos, brushtail and ringtail possums, and wallabies both 'fair to large' and 'small and short'.)

Quolls may no longer be shot and skinned and sold at market, either in Tasmania or on the mainland, but they're still under threat – just not quite so directly. As on the mainland, a large-scale threat to quolls and other wildlife in Tasmania is fire. Just like in the tropical rainforests of Queensland, fires now regularly threaten Tasmanian landscapes that have never before endured anything like them. In January 2016 bushfires burned more than 72,000 hectares in Tasmania, including in Cradle Mountain–Lake St Clair National Park, and in the Walls of Jerusalem National Park. David Bowman of the University of Tasmania, writing in *The Conversation* only two weeks after the fires started burning, explained that these fires were especially devastating because they burned both temperate rainforest and alpine vegetation unique to Tasmania – the latter being a particular concern:

> It is extremely unlikely burnt areas with the endemic alpine flora will ever fully recover given the slow growth of these species and the increased risk of subsequent fires given the change to more flammable vegetation and the slow accumulation of peat soils, which takes thousands of years.
>
> Past fires have resulted in a permanent switch from the unique Tasmanian alpine vegetation to more fire-tolerant vegetation.

Bowman added: 'There are two ways to think about the recent fire situation in Tasmania. We can focus on the extreme climate conditions and unusual fire behaviour, or we can see what is happening as entirely predictable and consistent with climate change.'

His view was borne out only a few years later. From December 2018 to March 2019 a huge area of Tasmania burned. An independent report into these fires drew explicit parallels with 2016: 'There are two stories to tell about the impacts of these fires: the significant success of the Tasmanian fire agencies in protecting human life and property, and, as in 2016, the damage done to ecological and wilderness values in the Tasmanian Wilderness World Heritage Area.' The report makes clear the scale of the fires: 210,311 hectares were burned, and the perimeter of the burned area was nearly two thousand kilometres – making the fires of 2018–19 the largest in Tasmania for half a century.

Laying out the climactic conditions in the lead-up to the fires, the report tells a by-now familiar story:

> The summer of 2018–19 was Tasmania's second-warmest on record ... Most areas had little if any rain for about six weeks from late December to early February ... Vegetation that would ordinarily be too moist to burn became available fuel.
>
> Extensive bands of lightning extended across the western and southwestern areas of the State on 14 and 15 January, with the second event producing over 2400 dry lightning strikes. The absence of any associated rainfall and hot, dry conditions resulted in over 70 new fires breaking out across the State. Several of these became significant fires.

It wouldn't take long for the story to retell itself: in 2019–20 fires again burned in Tasmania, as they did on the mainland, though they weren't as severe. That's what we hope for, now: fires, but not as severe. There's no shortage of people willing to point out that

Australia has always burned – not for nothing do we routinely talk about 'fire season' – but the ease with which fires ignite now, the readiness of the bush to burn, dried to tinder from coastal heath to mountain rainforest, and the frequency of storms with no rain but only lightning, is something new. I remember one night, many years ago, as a teenager sitting with my parents outside a house they owned at the time near Cobargo on the far south coast of NSW – a tiny town that would become famous throughout the country after it was devastated in the 2019–20 fires and its residents angrily confronted visiting Prime Minister Scott Morrison, a man who once chuckled as he tossed a carefully laminated chunk of coal around the House of Representatives. On that night, my parents and I watched as a storm rolled across the mountains towards us, the rain obliterating one row of peaks after another. Lightning rolled with it, and we gasped as – just on the other side of the valley – one bolt struck a tree and for a few moments the tree glowed fiery orange, until the heavy rain riding in the lightning's wake extinguished it just as quickly as it had been lit. Nowadays we get the lightning, but the rain is much less common.

I ask David Hamilton how quolls in Tasmania might cope with this new age of mega-fires. He explains that none of his monitoring sites are in fire-affected areas, so he doesn't have data on this particular question but:

> I would guess that they would behave like most carnivores post-fire though – so immediately afterwards they might come in and scavenge around the fringes, but wouldn't really start frequenting affected areas until there was a reasonable amount of understorey regrowth...
>
> I would guess that further extensive fires over large areas (particularly those not normally prone to fire) would have a negative impact on population numbers.

He stresses to me, though, that this is just his conjecturing at the moment. Much as early European colonists must have been unable to imagine an Australian bush without abundant wildlife, even as they brought foxes and cats into that same bush and cleared huge areas of it, it's difficult to be certain how animals and the ecological communities they live in might adjust – or not – to the new age of fire and drought. Some species might adapt – but even the most adaptable species might struggle with rising temperatures, longer heatwaves, more extreme weather.

Katherine Tuft, general manager of a wildlife reserve in South Australia called Arid Recovery, tells me 'that is exactly the conversation we're having now in the last year, it's all we're talking about really'. Arid Recovery is 600 kilometres north of Adelaide, where my father grew up. In his book *Among the Pigeons*

John L. Read describes the creation of Arid Recovery in 1997. Rabbit haemorrhagic disease (or RHD), otherwise known as calicivirus, had been released in Australia the year before, and had an immediate impact:

> Warrens that had housed dozens of rabbits only weeks earlier, now lay barren, with only the sounds of blowflies buzzing around bloated carcasses. Once reptiles became inactive in the autumn, cats and foxes ran out of food. I had been studying the wildlife in the desert near Roxby Downs for over a decade and was stunned by the dramatic impact this rabbit virus had on the native and introduced animal populations. Long-term counts at several other research stations confirmed the lowest ever recorded numbers of cats and foxes. Soon after, rarely seen native rodents, marsupials and ground-nesting birds started to flourish. If there was ever a window of opportunity to create a safe haven for stick-nest rats and other vulnerable desert mammals, post-RHD 1997 was the time.

Read describes the concept as an 'exclosure': designed not to keep animals in, but to keep them out – cats in particular. 'The Arid Recovery fence was built according to the best design for a cat-proof fence known at the time, and all rabbits were removed. Without their key food source the last of the cats climbed out of their own accord.' A key part of the fence's design was a floppy top, designed to hang over and which would be unable to support a cat's weight. But early tests of the fence revealed its flaws: Read describes a tomcat, in an experiment, taking only thirty-six seconds to climb up the wooden fence post and leap over the floppy top, clearing it without having to touch it. So the founders of Arid Recovery clad the wooden posts in metal sheeting. This deterred a large cat but not a smaller one, which they tested next and which proved to be more agile. Through this system of trial and error a final design was eventually arrived at: metal-clad posts, a larger floppy top, two electric wires, plus finer netting at the base of the fence to prevent young rabbits getting through. 'Subsequent expansions to the reserve were built with only two wider mesh strips,' Read writes, 'joined at knee-height and supported on recycled metal-bore casing posts that cats couldn't climb. This modified fence soon became the prototype for cat-proof sanctuaries springing up elsewhere around the world.'

It took two years for the vegetation within the fenced exclosure to regrow to the point where Read and his colleagues, Peter Copley and Katherine Moseby, felt comfortable reintroducing the first native mammals: greater stick-nest rats, once found from western NSW across inland Australia to the coast of Western Australia but by 1997 found only on the tiny Franklin Islands 20 kilometres off the coast of South Australia. Since then other species have been reintroduced, including the iconic greater bilby – and the western quoll.

When I talk to Katherine Tuft on the phone one Friday evening in late February 2020, she's setting out traps for one of those quolls, a first-year female, who isn't where she's supposed to be: the reserve, Tuft explains, is divided into six paddocks, and one of them is set aside to be quoll-free, 'so we have an experimental control and also as an insurance area just in case the quolls eat too many of the other animals' – and the quoll has got into it. She's a 2-kilogram needle in a 14-square-kilometre haystack, and she's mobile.

Tuft tells me that the reintroduction of quolls to Arid Recovery is still in its early days: twelve quolls – eight females and four males – were released in the reserve in 2018, after a trial run in 2015–16. In 2019 the reserve saw six of those females give birth to 32 young: making a total – with the males, and the female quolls that didn't breed – of 44 western quolls, inhabiting 123 square kilometres of the vast land their forebears once thrived in.

As in any animal population, genetic diversity is important to keep the quolls at Arid Recovery fit and healthy. To some extent the quolls take care of this themselves: 'northern quolls ... can have multiple fathers but in the one litter,' Tuft tells me, and through studying the quolls at Arid Recovery Tuft and her colleagues were able to determine that the same thing occurs in western quolls. Nonetheless, in 2019 'we had an honours student [Tessa Manning] do some work looking at the paternity of the first generation of quolls born here and she found that one male contributed to about sixty percent of the offspring'. Fortunately there's another reintroduction programme in Ikara-Flinders Ranges National Park, only five hours' drive away – no time at all in inland Australia – and Arid Recovery foresees being able to swap animals between the two projects, to increase genetic diversity in both locations. Although two of Arid Recovery's founder quolls, Tuft tells me, came from the Flinders Ranges, ten of them came from the wild Western Australian population, ensuring a maximum of genetic diversity within the two newly re-established South Australian quoll populations.

Cats are the biggest threats to quolls in inland South Australia – and to other small-and medium-sized mammals. Foxes are present in the area, Tuft explains, but 'cats are more common ... We've got a database of two thousand cats that have been dissected [since the reserve was founded] and it's clear that the larger cats, 3.5 kilos or more, are the ones that kill larger animals like rabbits, bilbies, bettongs, and quolls.'

Cats – feral and domestic – are now ubiquitous in Australia. There are perhaps as many as six million cats spread across 99.8 per cent of the continent, and predation by cats has already caused the extinction of 20 Australian mammal species. Efforts are continually underway to cull feral cat numbers in Australia, but with such a vast number of cats other approaches need to be considered. At the same time, a fenced reserve – even a large one like Arid Recovery – will at some point reach the limit of how many animals it can sustain, particularly given the nature of the environment in inland Australia. Tuft tells me the carrying capacity of the reserve – how many quolls it can support – fluctuates according to the boom-and-bust cycle of an arid environment, where prey can be abundant or nearly absent from year to year: 'we anticipate that there's going to come a point, and we've reached it already, where the reserve isn't big enough for the number of quolls'.

As shown by Tuft's attempts to catch the rogue quoll, the design of Arid Recovery is more complex than just a large fence around an area of bush. Each of the reserve's six sections has its own purpose. The First Expansion and Second Expansion in the reserve's south, and the Northern Expansion above them, contain all the species that have been reintroduced, and are free of feral animals. The Main Exclosure next to them is like the First and Second Expansions, except for the absence of quolls. But the two northernmost sections of the reserve, called the Red Lake Expansion and the Dingo Paddock, are where the most intriguing research takes place.

They're 'working towards essentially using the reserve as it is, particularly with its separate compartments, to test ways that animals might be able to survive outside,' Tuft says. 'A lot of that comes from recognising that this might be one of the largest reserves of its kind ... it's 123 square kilometres, but it's still pretty small in the scheme of things and it's very expensive to maintain ... it's not a forever solution.' She explains the longer-term goals to me a little more:

> [W]e've got to find ways for [the animals] to survive outside and that will include better ways of managing with feral predators that threaten them, but also making the animals themselves more resilient. So that's one of the other projects we're working on, trying to train animals to live with cats ... we've had a project running on that subject for [four or five] years now ... And we've seen behavioural change which is really cool...

Tuft tells me that the project was the brainchild of Katherine Moseby, who's Arid Recovery's principal scientist. Elaborating on the behavioural changes, Tuft says:

> [W]e had a population of bettongs and bilbies living with a controlled number of feral cats and they changed their behaviours and when we've moved bilbies that lived with cats with another group of bilbies that are naive to cats the bilbies that have cat experience survive much better...

Tuft sounds cautiously optimistic, but excited about these results: '[T]here's a bit of hope there but there's definitely a lot more work to be done,' she says. '[B]ettongs and bilbies will encounter cats over time and some will get eaten, [but] hopefully the ones that are left are [better] equipped for living with cats than the ones that get eaten.' It's a dynamic kind of conservation:

> [W]e're trying to kick along natural selection and speed up evolution ... hopefully there'll be some learning in there soon ... we'll find out in the next few years how much of that is natural selection versus learning and whether ... the mothers actually pass on that knowledge to the young or if [the young] have to learn it on their own.

For now, the study necessitates keeping quolls out – the presence in the study site of another predator other than cats will only complicate the research – but in the future Tuft and her colleagues hope to be able to compare the population-dynamics of prey animals in three situations: being preyed upon by cats, being preyed upon by quolls, or not being preyed upon at all. As Tuft points out, 'it'd be nice to be able to do this predator training with a native predator rather than with cats'. Not that the quolls are in the reserve only so that prey animals can learn how to avoid them: 'we had way too many bettongs for a good while there,' Tuft says, 'and bringing in quolls was one of the things we were trying to do to address that ... having herbivores in a confined area where they didn't have sufficient predation pressure meant they just ballooned out of control' – to the detriment of some of the other species in the reserve.

As for the quolls, they may have to learn how to live alongside cats, too. The limited number of quolls that the reserve can support 'presents an opportunity to establish them in the surrounding landscape if they can move in and out', Tuft says. There's still a lot to learn about the extent to which quolls disperse out of the reserve and how well they survive beyond the reserve's fences, but 'we'd love to see any of these threatened species surviving without a fence, and quolls are probably a decent one to aim for because they do persist in WA without a fence, so it can be done'.

The big question is under what conditions it can be done. And the conditions are about more than just the presence or absence of cats. '[T]he reintroduction of the quolls coincided with the onset of drought here,' Tuft tells me. This has affected the quolls' breeding in the two seasons they've been in Arid Recovery to February 2020:

> [T]he first [breeding season] was relatively late compared to other populations around the country and this year [2019–20] was the latest ever recorded anywhere, which we suspect was linked to the drought ... we were quite concerned that that would mean that the mothers would be raising their young at the hottest time of year [because] they were at peak lactational demand

> in late November and December [2019] when it was getting really, really hot.

Tuft and her colleagues were able to provide the quolls with extra water and food during this period: 'we were picking roadkill kangaroos off the road and dumping them in some spots for the quolls to chow down'. But the question remains of how the quolls – and the other animals at Arid Recovery – will cope in years to come. Australia is no longer looking ahead at climate change in the future but is living in it, now, and the extreme weather patterns created by climate change will only get worse. In my lifetime I've lived through two record-breaking droughts; my memories of the months of the so-called Black Summer are still fresh, as are my memories of Black Saturday before, and the 2003 Canberra bushfires before then. I've become accustomed to seeing Merri Creek at the end of my inner-Melbourne street in flash flood, the bike path that follows it regularly inundated.

Tuft tells me that the recent drought at Arid Recovery was unprecedented in the reserve's history – and not just because of the lack of rain but because of the extreme heat, too: '[There have been] some excessive periods of heatwave temperatures where we've had 40 degrees or more in a row repeatedly.' This seems to have affected stick-nest rats particularly badly: while the reserve's burrowing animals can hide from the heat underground, stick-nest rats live above ground. 'It's possible that we've actually lost them from the reserve this summer,' Tuft says. 'We're still trying to find signs of life.' For the stick-nest rats, the drought just added to the stresses created by predation from quolls in parts of the reserve that the two species cohabited in, and these stresses were exacerbated by the vegetation not being as resilient as it might otherwise be at other times 'because we had too many bettongs.' It's a lot to balance, and Tuft concludes:

> [T]here's potential to address these things and that's what we're thinking about now: how can we build up the resilience in the system to weather more droughts like this? Because there will be more. And what [species] do have a chance and what [species] do we say 'All right, we'll leave that to the temperate zone and we'll concentrate on species [that can survive the conditions].

***

As it happens, when I talk to Tuft on the phone, I'm far away from inland South Australia: I'm in Homebush, in Sydney, for twenty-four hours, to give a talk. On the train up to Sydney earlier that day, after a similarly brief trip home to Canberra, I'd suddenly remembered what Homebush had become famous for, apart from the 2000 Sydney Olympics – and, in a way, *because* of them: during the construction of the Olympic site, a population of endangered green and golden bell frogs had been discovered. The frogs are still there today, along with many other species and a complex of wetlands: travelling up to Sydney, I researched on my phone the best places to go in Homebush to try to hear them; which parts of the massive Homebush development had been set aside for them. And as I did so, the train passed through an area burned in the Green Wattle Creek fire, which in two months, from 29 November 2019 to 30 January 2020, burned 278,722 hectares. Passing between Bargo and Tahmoor a month after the fire had finally been contained, the black trees were fuzzy with bright new regrowth, as if hung with lichen.

In 2020 we all became fatigued by the word 'unprecedented'. The fires were 'unprecedented'; the drought that led up to them was 'unprecedented'; the floods that followed them were 'unprecedented'; the global pandemic that came hard on their heels was 'unprecedented'. In September 2015 then-prime minister Malcolm Turnbull spoke of the need for Australians to be 'agile'; but he was talking only about work, and fuelling the economy, not about the basic survival skills – physical and mental, emotional and social – needed just to live in a world fundamentally destabilised, without a centre, because of our almost obsessive need to take everything from it that we can get.

In May 2019, in an article for the ANU's College of Science, titled simply, 'How do we go on?', Tabitha

Carvan asked a number of researchers at the university exactly that question: 'how to exist in the face of existential dread'? 'It's not a matter of life will be really nice, and then we won't be here any more,' Carvan quotes Liz Hanna from the ANU's Climate Change Institute:

> It will be a bubbling up of all manner of things. People will be dying from heat, roads will melt, trainlines will buckle and we won't be able to get to and from where we need to go. There will be a shortage of water, problems with food supplies, blackouts.

Other interviewees in the article list a catalogue of other disasters, including, presciently, 'increase[d] contact between humans and animals', leading to 'a much greater chance of bugs spreading between them'.

A month into a six-week, stage-four lockdown in Melbourne, imposed to halt the spread of the coronavirus, a wind storm caused such damage to infrastructure that residents of eighty-eight Melbourne suburbs were advised to boil their tap water. A popular meme doing the rounds on social media in 2020 said: 'I could really go for some precedented times right now' – but also on social media, I see someone point out in the blur of daily posts that it makes more sense to think of 2020 – or any other coming year you care to think of – as not the hottest year in the last one hundred years but the coolest year in the next one hundred years.

We're living in the precedent, right now. Climate change has very quickly gone from being, in our minds, a thing that is going to happen, to being a thing that *is happening.* As Meehan Crist, writing in the *London Review of Books* in early March 2020, put it: 'We know enough to know that people are living with the effects of global heating right now.' But as Crist also points out, 'We know that climate risk and the worst effects of ecological disaster are unevenly distributed across race, class and gender, and among industrialised and developing countries.' But the way we talk about climate change makes it easy to believe that it's the cause of all our problems, rather than a symptom or an inevitable end-product. As former Australian Government Climate Commissioner Professor Will Steffen, quoted in Carvan's ANU article, points out:

> The problem we're facing, it's not a technology problem. It's not a science problem. It's a problem with the human side of things, with our socio-political values ... The neoliberal economic system we've bought into is completely at odds with how the Earth works. We have to change this value system that we operate under.

Carvan goes on to write:

> I used to feel like I was doing something. I was switching over my lightbulbs, cutting meat out of my diet, catching public transport, and printing on both sides of the paper. I felt virtuous. But

> now it's clear that these small acts of good citizenry are not working fast enough to make enough of an impact.

Crist, as have many others, explicitly criticises this model of climate action, which places the emphasis on individual action rather than systemic change:

> [S]hift[ing] responsibility for global emissions from systemic actors like fossil fuel companies and governments onto individuals ... gives corporations a pass while placing moral responsibility on people who live within systems where they are not free to make carbon-neutral choices. It accepts as inevitable the neoliberal order that has driven the climate crisis, and insists that our responses to this crisis take place within the same system. You can reduce your carbon footprint by eating less meat, but you can't buy carbon-neutral food. If you live in a city with poor public transport, you may have to drive to work.

Carvan writes at the start of her article that it was prompted, in part, by conversations with her young son:

> [He] was three and deep into his dinosaur phase. 'Will humans go extinct too?' he would ask ... 'Will the world ever end?' ... [He's] now six, [and] asking again if humans will go extinct ... And I say, 'Maybe, but not for a long time,' but really

> I'm thinking: please don't let it start until the generation after his.

Crist's article is also concerned with children, and the raising of them in an uncertain world: it's titled, simply, 'Is it OK to have a child?' But as she points out, this is a more complicated – and more troubling – question than it seems.

> Some people say they don't want to bring a child into a devastated world, or they don't want to harm the planet, or don't want their child to suffer in a degraded future. And then there is the moral imperative to erase that carbon footprint, which has as much to do with care for the environment and other people as with relieving the misplaced burden of moral responsibility...

In an increasingly complex and divided world, even such a primal question as whether to have a child or not is no longer as simple as it seems:

> Recently, the logic of the carbon footprint has become tangled up with ideas about global human population in alarming ways ... Listen closely to rights-based strategies to reduce carbon emissions through increased access to contraception and family planning. These strategies almost always involve black and brown women in developing countries having fewer babies. There is, of course, an unmet need for reproductive care and birth control in these countries, but we should be

> deeply sceptical of climate solutions that place the burden of solving the problem on women's bodies, particularly the bodies of poor black and brown women, while demanding very little of those who actually caused the problem. Such approaches aren't just ahistorical; in terms of global emissions they make no practical sense. Developing nations generally have much lower emissions per capita than the countries where such proposals tend to originate ...

According to Crist, calls to control the human population of the planet over-simplify the highly complex and poorly understood system of relationships between people, their institutions and the environment. As a result:

> There are no 'good' programmes for global population control, not just because it is anti-feminist, racist and anti-human, but because it is impossible to put an upper limit on how many people are too many for the planet ... it isn't just the total number of humans that matters, but the way humans organise to use the available resources. Today, by most metrics, we're doing terribly. The systems we have are utterly incapable of meeting the needs of the global population.

In our focus on mining, on forestry, on over-fishing – on any of the destructive practices entrenched in our global socioeconomic model – it's easy for us to overlook the fact that the first and fundamental extractive industry of capitalism and the colonialist mindset with which it goes hand-in-hand is the extraction of human labour and quality of life: that millions upon millions of people all over the world have given and continue to give their lives and health and safety and happiness to a system that has no interest in ever rewarding them for it, and which in fact cannot function without using those lives as grist to the mill. But when the responsibility for changing this system is put on individuals, we begin to view humans as something somehow separate from the natural world.

It's become almost reflexive for many people, including many who'd consider themselves progressive or kind or generous, to describe humans as a plague upon the earth, and to say that the world would be better off without us. This raises a question of what would become of those species which depend on us – from domestic dogs to the microscopic mites which live on our skin or the bacteria in our digestive systems – but it also doubles down on what's brought us to this age of disaster in the first place: a belief that we're somehow apart from it all, that we can ruin the natural world and not have to bear the consequences of it. Viewing humanity as a parasite upon the earth and viewing the earth as no more than a pool of resources from which to draw whatever we need are two sides of the same coin – and neither will get us out of this mess.

It's no coincidence that in this moment in history we're seeing a global resurgence of fascism and totalitarianism. When our entire way of life is predicated on the supposed virtue of making as much money as possible, for many people it can be easier and more comforting by far to believe that things are bad because of a vast undifferentiated mass of 'others' all scrapping for the same meagre resources than because an entrenched power system needs to maintain the status quo even if it's killing us all. It's seemed apparent for a long time that, regardless of party affiliation, a political system funded by neoliberalism and wholly invested in it will sooner shift the blame to supposed outsiders than agitate for any actual systemic change: the powerful will more readily consolidate power than share it.

In 2018 the Intergovernmental Panel on Climate Change announced that humanity had only twelve years left to keep global warming to a maximum of 1.5 degrees Celsius. The challenges faced by Australia are faced by the whole world. In 2019 a UN report based on a review of more than 15,000 studies – the largest such review ever conducted – declared that one million species worldwide were at risk of extinction. The number is so large as to be almost numbing: like astronomy, such a large figure just doesn't compute. Years ago when I was studying biology at university I was taught that the maximum number of objects that a human can look at, and instantly quantify without having to count, was five:

any more than five fingers, or five berries, or five family members, and we have to stop and consciously count it out. Our brains didn't evolve to understand things in the millions; such numbers are impossible to make sense of. The scale of the disaster we've created, dragging every living thing in with us, is unfathomable.

But it sinks in, slowly, like gradual rain seeping into the ground – an increasingly rare phenomenon in these days of drought and flood, these days of extremes. As that awful number gnawed at the back of my mind over days, over weeks, over months, I found myself thinking, seriously and undeniably for the first time, even as the urgent desire and realisation that I was emotionally ready to have a child grew simultaneously within me: what right do I have to bring a human life into a world that is rapidly being killed? What kind of a world would I be bringing another human life into; what kind of a world is being made or unmade for them? If I had a child, and raised that child to value the natural world as I was raised by my own parents, and if that child one day asked me why I had fathered them, and why I raised them to love things that were vanishing, never to be seen again, what could I possibly say to my child to justify my decision? What could my answer be? At the end of 2019, in the middle of the bushfire crisis, before the pandemic, I put a question out on social media to friends and strangers alike: have worries about the future of the environment and the planet affected your thoughts about whether or not to have children?

'Re having kids or not,' someone replied, 'I was always on the border, thinking maybe yes or maybe no, depending on who I met, but climate change and all the environmental disasters and depressing environmental issues have tipped me over the edge into "no, not having kids".' A friend, Lisa, tells me simply: 'I just can't imagine bringing a child into this world which is literally on fire.' The question of whether to have children or not in this world is often framed as being about what impact those children – another human life to sustain – will have on the world, but the more difficult and unsettling question of what impact the world will have on the children seems to be less often remarked upon.

It's an animal urge, to have offspring. Not everybody feels it, but for those who do it claws at them. I feel it myself: when I see a child my heart aches with yearning. I even get clucky when I see young magpies begging for food from their parents. I feel myself ready. I worry about how old I'll be when my child is an adult; whether my child will remember his or her grandparents; whether my parents will be too old to experience the full joy of being grandparents. As I get older my timeline for starting on family life shifts back ever further: from thirty, to mid-thirties, to forty – now to early forties. Not too late, surely.

On New Year's Eve 2019, Australian prime minister Scott Morrison said in a video address to the nation that there was 'no better place to raise kids anywhere on the planet'. At the same moment, the world watched in horror as 4000 people were evacuated to the edge of the water in Mallacoota, in far eastern Victoria: the defining image of the evacuation was of a child steering the outboard motor of his family's boat beyond the flames, out into the sea, a mask obscuring his young face, the whole frame of the photo – taken by his mother – washed in the red of fire and smoke. Exactly a month later, on 31 January, the Fact Check service jointly run by the Australian Broadcasting Corporation and the Royal Melbourne Institute of Technology stated that the oft-repeated figure of one billion animals having died as a result of the Australian bushfires of 2019–20 was a 'conservative estimate'.

Nowadays, suddenly, everybody who spends time thinking about the state of the environment is familiar with the concept of 'shifting baseline syndrome', that incremental downward creep in species number and diversity, step by step, so gradual that it isn't noticed. But we live now in a time when the baseline isn't shifting so much as sliding chaotically out of control, like a melting wall of glacier ice shearing off into the ocean. We talk to each other about how few insects there are now, compared to even a decade ago. A joke I learned when I was a child – unthinkingly callous as so many children's jokes are – goes like this: 'What's the last thing that goes through a fly's mind when it hits a car windscreen? Its bum.' It's a joke that is now almost absent of meaning, as elusive as the world's oldest known joke, from ancient Sumer nearly 4000 years ago: 'Something which has never occurred since time immemorial; a young woman did not fart in her husband's lap.' It strikes me as strange that we should know of such ancient civilisations, and their long-vanished customs, and yet think of our own current civilisation and our own way of living as eternal.

When the great American writer and thinker Ursula K. Le Guin died in January 2018 one quote was repeated again and again in obituaries written for her: 'We live in capitalism. Its power seems inescapable. So did the divine right of kings.' It's a salient, and necessary, reminder of the difficulty of seeing a way out of something that has enveloped us. Le Guin's mention of kings is pertinent: human civilisation has seen kingdom after kingdom, empire after empire, fall away, each one its own particular way of life that was all-encompassing for the people living within and under it. Capitalism now behaves as an empire, too, and in its global scale it must surely be the most powerful empire ever to have existed. Its scope and power, and the power of those who drive it on, are overwhelming – and they're overwhelming the entire planet, every ecosystem and species. But all the same, it's a human power. And Le Guin's quote continues, with hope and promise: 'Any human power can be resisted and changed by human beings.'

3

If you want to you can go online and find videos from Mulligans Flat in Canberra of baby eastern quolls frolicking round their natal dens. When a biologist is asked what the meaning of life is they're likely to say that it's to replicate, to reproduce. That's what I was taught when I was at university. And as reductive as that may sound, quolls exemplify this argument better than many other mammals: for such a relatively large mammal they live astonishingly short lives – only a few years. At a year old they're sexually mature: they can have offspring. They live fast lives, with high population turnover: male eastern quolls live to around two to three years, while females can live up to four years. After reaching maturity a female eastern quoll can raise up to half-a-dozen young each year.

The young quolls in the videos from Mulligans Flat are as playful as kittens – and for the same reason: they're predators. All that playful leaping and biting and investigating of nooks and crannies is developing skills and techniques that will keep them alive in adult life.

Quolls would have been plentiful around Mulligans Flat in 1857 when land in the area – perhaps named after a now-forgotten convict – was starting to be subdivided and sold. By that time, my ancestor Michael Farrell was a grandfather, and had been a free man for three decades, making a good life for himself as a victualler – a licensed seller of alcohol – and proprietor of a series of Sydney pubs: the Whale Fishery on Cambridge Street in the Rocks; the Welsh Harp on George Street in the city; the Farriers Arms, also on George Street. In the very same year that the land at Mulligans Flat was being sold, 1857, the last known specimen of Gould's mouse was collected in country to the west of Sydney. In the previous decade tantalising fragments of the bodies of two hopping mice, the big-eared hopping mouse and the Darling Downs hopping mouse, had been collected on opposite sides of the Australian continent – the first and last time these species had been noted by Western science, an enigmatic start to colonial Australia's spiral of mammal extinctions. Today, if you visit the airport in Canberra, as soon as you arrive you can play a game in which you spotlight for virtual representations of various animals that have been reintroduced to Mulligans Flat – and if you're visiting during the warmer months, or if you live in Canberra, you can even sign up for the reserve's seasonal 'turtle patrol'. In August the eastern long-necked turtles that live in the dams at Mulligans Flat start migrating across land, and the large fence, like the one at Arid Recovery, which runs around the heart of the reserve – which was completed in 2009, at a length of 11.5 kilometres – can present an obstacle. A blog post on Mulligans Flat's website from 2016 notes that: 'Last year more than 160 people registered their interest in helping us patrol the predator-proof fence'.

Belinda Wilson was once a turtle, patrol volunteer. I first met her in September 2018, on one of my return visits to Canberra, when my parents and I had joined one of Mulligans Flat's regular twilight tours and she had been the guide. In February 2020 when I meet her again she's at the tail-end of her PhD on the reintroduction of eastern quolls to Mulligans Flat, and she allows me to accompany her as she sets up cage traps for one of two scheduled eastern quoll surveys in the reserve for the year. We spend two hours driving what looks on the map that she hands me like a simple loop, but in which I become thoroughly disoriented, and we stop every couple of hundred metres to re-bait a trap the location of which she has memorised through repetition.

Wilson tells me that eastern quolls were reintroduced to Mulligans Flat in a series of four releases, from 2016 to 2019. There were approximately forty-seven quolls in total, most of them female. By the time the quolls arrived, eastern bettongs, New Holland mice, and bush-stone curlews had already been reintroduced to the reserve – in 2012, 2013 and 2014 respectively: all of them species that had once thrived in the Canberra region and had long since been hunted to extinction by introduced predators or died due to habitat clearance.

The fence at Mulligans Flat is designed to keep out both foxes and cats, but control efforts don't stop at the reserve's border: Mulligans Flat is flanked on its entire western side by the suburbs of Gungahlin, the northernmost of the various satellite towns that together make up Canberra. All the suburbs adjacent to Mulligans Flat – Jacka, Bonner, Forde – are designated as 'cat containment' suburbs, with rangers able to seize cats found outside private property and the cat's owner liable to be fined up to $1500 for the infringement. A distinctive cat-containment symbol – literally, a cat inside a house – appears on street signs throughout the suburbs to remind residents.

Wilson tells me that early residents of this relatively new part of Canberra all bought in to the idea of cat containment, but that more recent residents have been more stubborn. It's just one of the many challenges faced by the animals in the reserve – as the Mulligans Flat Draft Strategy document released in 2020 puts it, the reserve's ongoing aim looking ahead to 2045 is 'rebuilding natural and cultural values, systems and processes to become more resilient despite complex threatening processes'.

Community buy-in is important, not just because of the proximity of the reserve to Canberra's northern suburbs but also because of its design: the fence keeps animals out, but it's not designed to keep them in. Already some animals from the reserve occasionally turn up in the suburbs outside the fence, and if everything goes to plan then by 2045, the Draft Strategy says, 'reintroduced native fauna [will] have established themselves beyond the Sanctuary's fox and cat proof fence, with strong support from local and regional communities' and '[l]inking with the landscape [will have] inspired changes to ecosystem management beyond the fence, bringing a welcome diversity into our landscapes.'

Eastern quolls are one of the animals that sometimes go over the fence in the present day. Male quolls tend to wander more than females, particularly in the breeding season. In fact one of the obstacles to reintroducing eastern quolls to Mulligans Flat in the first release in 2016 was that the males kept climbing out of the reserve. In response, Wilson changed the procedure for subsequent releases: female quolls were brought to the reserve after the breeding season, when they had fused pouch young. In this way, instead of adult males which dispersed over the fence as soon as they were released, Mulligans Flat built up a population of young male eastern quolls still in their mothers' pouches. After this change, survival levels increased as the reintroductions continued. Several years later the population appears to be self-sustaining: with a lifespan of only a few years, and as many as six young per female per litter, there's a high population turnover. Every year, as much as 90 per cent of the reserve's eastern quoll population is new. In the week that I visit the reserve with Wilson, across three nights of monitoring via trap-and-release, thirty-eight quolls are caught in total, with twenty of those being caught for the first time. A few months later, in May, Wilson and her team are due to conduct the second round of trapping, which will indicate the size of the reserve's breeding population. Wilson tells me that in February the quoll population in the reserve is around sixty; by breeding season it will be around forty to forty-five. She estimates that the reserve's carrying capacity is forty quolls, though she's quick to point out that she doesn't yet have the data to confirm that.

The reserve is bright green on the day I visit. After months of horror for south-eastern Australia in general and Canberra in particular a sudden burst of heavy rain has brought the reserve back to life. Wilson and her colleague Shoshana Rapley, who's studying the reintroduction of bushstone curlews to Mulligans Flat, comment often on how dry the reserve was just weeks earlier, as drought and bushfire and smoke assailed the ACT and surrounding countryside.

During the height of the bushfires of 2019–20, I arranged for a friend of a friend who was driving from Melbourne to Canberra to deliver a box of filtered masks to my parents, both of whom are in their seventies. They live now in one of Canberra's new suburbs, Wright, in the Molonglo Valley, 30 kilometres south-west of Mulligans Flat. Like the suburbs adjacent to Mulligans Flat, Wright is a cat-containment suburb. My parents don't have a cat but they have a dog, who they take for walks sometimes on the lower slopes of Mt Stromlo, just a few blocks from their house. There they usually pass by the Canberra Bushfire Memorial, built to commemorate the Canberra fires of 2003, which destroyed five hundred houses and killed four people, as well as burning up the slopes of Mt Stromlo and wreaking enormous damage upon the observatory at the mountain's peak. Where my parents live was once a pine plantation, of the type that are scattered throughout the Canberra suburbs and which fed the flames on that January day in 2003.

The catastrophe that day started when a dry lightning strike in Namadgi National Park sparked a fire that burned unchecked for two weeks before being blown into the city's south-western suburbs on fast, hot, gusty winds. Namadgi makes up the whole southern half of the ACT, and whenever I fly into Canberra the turbulence of flying over the mountains of Namadgi tells me that I'm home: the national park is precious to every Canberran.

Flying to Canberra is expensive at the best of times, though, so I only do it a few times a year, on set occasions. One of them is my birthday in early September. In September 2019, for my 40th birthday, my father and I went for a day-walk in Namadgi, up to Square Rock in Corin Forest, where we sat looking out over the valleys below. On the way back down we took a short detour to check out a lookout in the other direction, over the Orroral Valley, but trees had grown tall in front of the lookout, and we couldn't get a good view so we didn't linger. On the way up and back down the mountain, we walked past an area of extensive burning: we could see that it covered only the ground, and didn't extend for more than a couple of metres up the tree trunks, and had a clearly and sharply define edge – so we deduced that it was hazard-reduction work.

Bushfires were already burning in northern New South Wales at that time, hundreds of kilometres away, but we had no idea what was ahead: four months after our walk, on the afternoon of 27 January, heat from a landing light on the underside of a military helicopter undertaking bushfire-control preparation after two months of fires in the Canberra area ignited dry grass in the Orroral Valley. From this single spark, which took only 12 seconds to ignite, 82,700 hectares – 80 per cent of Namadgi National Park – burned, including the hazard-reduced forest that my father and I had walked through only months earlier.

When the Orroral Valley fire burned through Namadgi I saw photos of it on social media taken from the top of Mt Stromlo. My parents had their evacuation plan in place, just in case, and they and my brother, who is also in Canberra, lived in a state of high alert. It was a season of calling my family, time after time, to check that they were okay: okay in the midst of the smoke that came first from fires burning around Braidwood to the city's east and which temporarily made Canberra – clean, crisp-aired Canberra – the most polluted city in the world; then, okay after a cataclysmic hailstorm hit the city; then, okay during the long Orroral Valley fire. For months from the end of 2019 to the start of 2020, it felt as though we were living in the future – people throughout Australia took photos and commented on how they looked like they'd come from some dystopian science fiction film – and we wondered if this was what all our years would be like from now on. It feels as though a continent as vast as Australia can burn endlessly, if conditions allow it.

The drought and hot winds of the extended fire season seem a long way in the past on the day I meet Belinda Wilson at Mulligans Flat, though. It's easy to forget, for a moment, all that the country has just been through – Canberra more than any other capital city. But the subject always lurks. I ask Wilson how she and her colleagues handled those long months and she tells me that, in the midst of it, her PhD supervisor sent an email around to all his students enquiring after their wellbeing. She also tells me that, in her spare time, she volunteers monitoring the fruit-bat colony in Commonwealth Park, in the middle of Canberra on the northern shore of Lake Burley Griffin – a place that had a legendary allure when I was a child thanks largely to what my friends and I called 'the castle', a rock playground of tunnels, turrets and fortifications, perfect for a child's imagination. Fruit bats were unheard of in Canberra when I was growing up there, in the eighties and nineties – they were first recorded in the city in 2003, the same year that I left Canberra first for England and then for Melbourne. Now the population of the colony is as many as eight thousand, though a report in *The Canberra Times* from just before Christmas 2019 notes that that month's heatwave – 'unprecedented for the month of December' – was killing the bats: heatwaves kill fruit bats with increasing frequency wherever they're found in Australia. Only a month later, on 20 January 2020, a freak storm dropped 'golf-ball sized hail stones' on Canberra, including on Commonwealth Park. Wilson tells me of tending to the bat colony in the storm's aftermath, of wings with compound fractures and volunteer vets having to euthanise animals on the spot.

Wilson obviously cares deeply about wildlife. As we're driving through Mulligans Flat to the start of the trap line she spots a raptor circling high above a nearby ridge – we eventually conclude that it's a brown goshawk. During our time at Mulligans Flat she points out wallabies and echidnas – and the ubiquitous eastern grey kangaroos. The kangaroos are a bit of a problem: although native to the area, they've proliferated to such a degree – not just in Mulligans Flat but in Canberra in general – that Wilson tells me they form one of the biggest obstacles to ecosystem regeneration.

A section of the reserve within the main fence is further fenced to keep kangaroo numbers low – part of an ongoing monitoring programme to determine the effects on native grassy woodlands of overgrazing by kangaroos. Wilson points out the plot to me: the grass behind its fence is tall and lush. That grass, Wilson explains, provides shelter for the small ground-dwelling native mammals – bettongs, bandicoots, native mice – that have been reintroduced to the reserve. Wilson tells me that in order to restore the ecosystems within the reserve, just like those outside, it will take more than just quolls: other, larger predators are needed. I'm reminded of perhaps the most famous example of 'rewilding': the reintroduction of grey wolves into Yellowstone National Park in the US. According to the Wyoming writer Brodie Farquhar, after the eradication of wolves in the national park in the 1930s:

> Even though Yellowstone elk were still preyed upon by black and grizzly bears, cougars and, to a lesser extent, coyotes, the absence of wolves took a huge amount of predatory pressure off the elk ... As a result, elk populations did very well – perhaps too well. Two things happened: the elk pushed the limits of Yellowstone's carrying capacity, and they didn't move around much in the winter – browsing heavily on young willow, aspen and cottonwood plants. That was tough for beaver, who need willows to survive in winter.

The result of bringing wolves back has been a trophic cascade of healthier ecosystem effects: the wolves prey on the elk, and with fewer elk kept on the move by hungry wolves, riverside vegetation has been able to re-establish; this has created more food for beavers, as well as construction material, resulting in more beavers in Yellowstone and more beaver dams. Those dams, Farquhar writes, 'have multiple effects on stream hydrology. They even out the seasonal pulses of runoff; store water for recharging the water table; and provide cold, shaded water for fish, while the now robust willow stands provide habitat for songbirds.' All from reintroducing one apex predator.

Mainland Australia, though, is poor in large mammalian predators. The obvious species is the dingo, but Wilson explains to me that dingoes are fundamentally unsuitable for a reserve such as Mulligans Flat: they have a habit of herding prey animals into an area and killing them. Not so much of a problem in the wide, open spaces of wild Australia, but in a fenced area like Mulligans Flat – even such a large one – it could be disastrous. Restoring an Australian landscape after two centuries of enormous damage and upheaval is not as simple as just putting everything back and hoping for the best – it needs to be done with more care and consideration.

So the quolls at Mulligans Flat find themselves in a very different world from that which the Canberra region's original quolls would have inhabited. Wilson says that many of the burrows that the Mulligans Flat quolls sleep in during the day are old rabbit warrens. She also tells me that a biodiversity offset funded in response to the construction of a new suburb has allowed the fence at Mulligans Flat to be expanded to its sister reserve, Goorooyarroo, to the south. She's hopeful that quolls will self-release from Mulligans Flat into Goorooyarroo and prey on the rabbits there. There's precedent for this: in 1901 *The Manaro Mercury* newspaper reported on a 'conference of delegates from the Pastures and Stock Board of the Southern Districts' who met in Goulburn 'for the purpose of discussing the rabbit question', where among other suggestions the motion was carried that both eastern and spotted-tailed quolls should be protected as 'enemies of the rabbit'. In fact Wilson forwards me a paper published in 2013 in *Australian Journal of Zoology* by David Peacock, from Biosecurity SA, and Ian Abbott, from Western Australia's Department of Environment and Conservation, which claims that early attempts to introduce rabbits to Australia were unsuccessful in part because of predation by eastern quolls. And as suggested by Wilson's hoped-for quoll activities in Goorooyarroo, this effect isn't just historic. Peacock and Abbott note that:

> A proposal to reintroduce *Dasyurus geoffroii* [the western quoll] to South Australia's Flinders Ranges after their loss in the 19th century is being actively resourced and planned and primarily targets ameliorating the extreme negative impacts of low-abundance rabbits in arid fox-baited conservation areas. One of the anticipated benefits of this reintroduction is their direct and indirect impacts on low-density populations of European rabbits.

As we set each trap in turn, driving a hired van – not an ideal vehicle for the dirt roads in Mulligans Flat, Wilson says, but after the recent hailstorm, which destroyed the windscreens of the project vehicles, utes and four-wheel drives are in high demand and short supply in Canberra's hire-car industry – Wilson tells me about the early days of her trapping, figuring out by trial and error which bait was most effective in luring quolls. It turns out that the answer was tinned fish, and not, to her relief, rotting meat – though she tried that, too. Pointing out roof tiles placed on top of each long wire trap, she explains that the bettongs, essentially very small kangaroos, and nocturnal like the quolls, have a habit of jumping onto the traps, and the jolt of their small bodies can cause the trap to spring even though no animal is inside it and no quoll has taken the bait. The roof tiles act as shock absorbers, and before leaving each trap Wilson gives it a few good bangs to make sure that the door isn't going to fall shut accidentally.

While we're setting traps three other teams are out doing the same thing along different roads; in this week of monitoring, they have all been up at 2am that morning checking the traps and measuring each quoll caught; they also check the quolls for microchips or, if it hadn't been caught before, implant a microchip. They'll be up again at 2am tomorrow morning, to check the traps set today. Two nights in a week they do this, and before we head out to set the traps I overhear Wilson debriefing her team on the events of the night before, detailing how processes could be improved. I'm impressed, as always, by the dedication of people working to conserve Australian wildlife, the creativity and endurance to stretch scant resources to the ends needed.

Every now and then as we reset traps, Wilson finds a quoll scat inside one. It's as close as I get to seeing the quolls: the only evidence I have, apart from Wilson's word, that the animals exist. It's a hot day, and despite the resurgence of greenery after the recent rain, the bush still feels slow to shake the drought, and is strangely silent. While I'd been waiting for Wilson, near the entrance to the reserve among some pine trees, I heard and then saw a young superb parrot, a species that is listed by the federal government as vulnerable, and that breeds at Mulligans Flat – but other than that few birds make their presence known. I wonder sometimes how the bush would have appeared before the arrival of Europeans, and even to those alien eyes and ears upon their first arrival.

When I lived for half a year with my English grandparents in Surrey, in 2003, after I graduated from university I was surprised by the boldness of English mammals: how that country's rabbits, squirrels, deer, even foxes patrolled their territories during broad daylight. I was so accustomed to the nocturnal or crepuscular habits of Australian mammals that I assumed that must be the norm: although birds are ever-present here I wasn't used to seeing wild mammals going about their business during the day. Prior to two centuries of cats and foxes, of rabbits and sheep and cows, of grazing and clearing and slashing and burning and trapping, was Australian mammal life abundant and bold once the sun went down? Was the night as alive with the calls of quolls and other mammals as the daytime is with birdsong? Colonial activity has obliterated so many animals, but have we also changed the behaviour of those that survive?

One night I was walking home from Merri Creek at the end of my street in the inner north of Melbourne just after dark when, a few doors from my house, I saw something I'd never seen before: a common ringtail possum on the ground. Ringtails are called 'common' for a reason: there's one that lives in the wall of my house; I often hear them trilling to each other from people's gardens when I walk around at night, to the train station or the tram stop or the shops. They're so abundant in the suburbs of Melbourne that they even sustain a good number of powerful owls, Australia's largest owl species, more typically associated with deep, wet forests. I've seen more hundreds of ringtail possums in my life than I could possibly remember – but never, ever one on the ground.

This one was feeding on nuts dropped on the road by the ornamental hazel trees planted along the street. It's possible that ringtail possums do this every year, in this seasonal abundance, and I've just never noticed it before. But the sight struck me not just because of its novelty but because of the timing: it was 5 April 2020. The Victorian premier, Daniel Andrews, had declared a State of Emergency three weeks earlier. Only one week before I saw the possum, stage-three restrictions had been put in place, making people liable to be fined if they were outside for any reason but shopping for essentials, medical reasons, work or study, or exercise. Exercise was my excuse.

The reason for all this, of course, was the COVID-19 pandemic. At first we heard whispers, in vague but largely unconcerned headlines, of a new virus. By the time the news of Italy entering lockdown reached us, the virus – and the increasingly erratic responses to it from governments around the world – had spread, taking over our lives almost before we knew what was happening. Weeks seemed to be months; riding my bike to get groceries one day, just before stage-three, I passed a pub with a Valentine's Day display in its window, and I thought, 'Wow, that was a long time ago' – before realising that it had only been the previous month.

Shortly into the pandemic stories of animals reclaiming abandoned streets began appearing, often debunked only the day after going – in the now-ironic term used by the digital world – 'viral'. Soon the trend spawned a short-lived meme: people posting photos of rubbish or some other detritus of human civilisation accumulating on the streets and joking that 'wildlife is returning to the cities' – always good for a grim laugh in these bleak times. It was in this context, at this time, that I saw the ringtail possum on the ground, and I'll likely never know if it was exhibiting natural behaviour long suppressed by the presence of people, or if it was merely something that I'd never noticed before.

Replying to my social media question in 2019 about having children, one of my friends, James, told me: 'I've pretty much always known I don't want children and none of my partners do either. The changing climate and the rise of fascism really only cements that for me.'

His mention of fascism is pertinent. Although often well-intentioned, the posts and comments on social media during the pandemic about wildlife returning to cities, about the sudden cleanness of a world without traffic pollution, betrayed a naive thoughtlessness about the real impacts of the pandemic; not just that the virus directly and indirectly destroyed lives, but that those lives – whether uninsured Americans, immunocompromised people or the elderly, or even healthy people who lost income, jobs, homes, support networks through an increasing casualisation of the workforce and governments that fail to provide adequate safety nets – overwhelmingly belong to the most vulnerable and most disadvantaged in society. No good and equitable environmental movement can be built on the back of the entrenched power systems that have led us to this very moment of catastrophe: to celebrate a greener world yet ignore or even embrace or further enforce inequity is to run the risk of indulging in ecofascism. As US academics Jordan Dyett and Cassidy Thomas put it in a paper published in *Perspectives on Global Development and Technology* in 2019:

> While ignorance obviously does not qualify one as an ecofascist, it is vital to consider the consequences of the approaches and scenarios that this ignorance may precipitate. In refusing to address the root causes of various modern crises ... they will be allowed to reproduce and increase in intensity.

With the resurgence of far-right movements around the world, most strikingly in the US, it's no surprise that such attitudes are seeping into conversations about how nature might be 'healed'.

One of the more preposterous climate-change conspiracy theories is that scientists are perpetuating a climate change 'hoax' as a way to get rich off grant money for their research. Anybody who's ever met a scientist will know that it's an ascetic existence, and vulnerable to forces well beyond the control of the scientist. The day after I see the ringtail possum eating the nuts on my street I email Wilson to see how restrictions due to COVID-19 have affected her work. She replies the next day:

> COVID-19 has had a significant impact on our work, with social distancing restrictions causing the ANU Animal Ethics Committee to cancel all future fieldwork.... the quoll monitoring session that was scheduled for May cannot go ahead,

which will impact two ongoing studies of population dynamics that began in 2016.

The February monitoring session gives us an idea of how much juvenile recruitment contributed into the population after they become independent of their mothers in November. The May session, however, helps us find out who remains at the start of the breeding season which indicates effective population size (those that contribute to reproduction later in May). That announcement of fieldwork cancellation by the ANU also means that the fieldwork phase of my PhD has come to an abrupt end. While this is good in some ways and allows me to start writing up my results earlier, it is also sad that the February session will be the last quoll monitoring session I will run under my PhD. Apart from anything else, I will really miss the quolls...

***

Genetic diversity is crucial to the long-term viability of most populations of most species, including the eastern quolls at Flat. Eastern quolls were predated and persecuted to seeming extinction on mainland Australia, surviving in the wild only in Tasmania – so it's to Tasmania that the ancestry of all eastern quolls on the mainland can now be traced. Wilson explained to me that the quolls at Mulligans Flat were relocated from four different, mutually isolated populations in

Tasmania, in order to allow as much genetic diversity as possible in the new population. Quolls were also sourced from Mt Rothwell, a reserve located midway between Melbourne and Geelong and which had also relocated quolls from Tasmania.

I visited Mt Rothwell in April 2018 at the impetus of a friend. Like Mulligans Flat, the heart of Mt Rothwell is a fenced-off tract of remnant open grassy woodland and basalt grassland. And also like Mulligans Flat, Mt Rothwell offers monthly guided dusk tours: my friend Kate, as eager about wildlife and nature as I am, booked tickets for a group of her friends, including our mutual friend Katherine and me, to attend one of those tours.

We spent the afternoon first at the Western Treatment Plant, driving around the complex looking at the abundant birdlife in what's sometimes referred to as 'the Kakadu of the south'. The You Yangs loomed to the north. As dusk approached we left the treatment plant and drove towards them, over the Princes Freeway and past the small town of Little River, through farmland: the grassy plains here, like grassy plains everywhere, were readily consumed by agricultural and urban development, a process continuing even today as Melbourne expands west into the degraded landscape. Species aren't the only things that can be endangered: the ecosystems in which they live can be, too, and Victoria's volcanic grassland is a critically endangered ecosystem. A 2011 report from what was then Australia's Department of Sustainability,

Environment, Water, Population and Communities stated that less than 5 per cent of the habitat's original expanse was left; according to information on Mt Rothwell's website, that number is now less than 1 per cent.

Mt Rothwell protects 473 hectares of this grassland and its associated grassy woodland – also a critically endangered ecosystem. As my friends and I drive towards the reserve, scores of black kites and whistling kites settle into the trees around us. Several other people, including families with young children, have also booked their place in the evening's tour, and as we all wait in our cars a guide arrives on a quad bike and lets us through the gate. We gather under the trees outside some low buildings while we wait for the tour to start; it's April, that time of year when the night air starts to gather an edge of cold, and the sky is clear, and as the sun drops I start to shiver: away from the urban heat of Melbourne the temperature is lower than I'd anticipated. Kate lends me a jacket and I gratefully accept.

Two dingoes lounge in a small enclosure near the buildings, leather collars around their necks. I've never seen a dingo before: they're a threatened species in Victoria, and nominally protected – though their status reverts to mere 'wild dog' when they're deemed to be a threat to livestock, and as such they can legally be 'controlled'. Quolls, similarly, were once regarded as a threat to poultry, and so were heavily persecuted up until the 20th century. When the tour starts, after we've all had time to get dinner, and the dusk has become dark enough to prompt the emergence of the reserve's nocturnal residents, we're led first past a small building where breeding eastern quolls are kept before release. A barn owl circles above us, screeching and settling into the top of a dead tree, the bird just a white smudge against the light pollution that spills into the night sky from distant Melbourne.

We walk up a dirt road, stopping from time to time to listen for bats. Here, as everywhere in southern Australia, white-striped freetail bats fly fast and high above us in the night sky, one of the few microbats whose echolocation calls are within human hearing range. Our guide describes the call as like two 20c coins striking each other – though they're mostly silent tonight. Perhaps in the cold air there are few insects about, and the bats are in torpor in the crevices and hollows that they sleep in. Our guide sweeps a red torch-beam across the ground, looking for eye-shine – red light is less disruptive to the reserve's animals than white light – and tells us about Mt Rothwell: how the fence keeps cats and foxes out, but they're still in the process of eradicating rabbits within the sanctuary; how the fence was once designed to allow animals to leave the reserve and repopulate the countryside beyond, but that it had to be redesigned after it was found that every animal which did so perished, either falling victim to the cats and foxes still abundant beyond the fence, or killed by cars on the dark dirt roads which crisscross the farmland around Mt Rothwell. Now the fence keeps animals in as well as out, and it's patrolled daily, all year round, to ensure that there's no damage and that no rabbits have burrowed under it. Any weakness or breach of the fence must be repaired without hesitation.

It's ongoing work, this stewardship of the natural world – in whatever form it takes. Without human intervention the world may reach some kind of equilibrium but, with all that we've done so far to upset the balance, the resettled world may be something that we don't recognise, something poorer by far than what it was before. I think again of that strangely empty bush in Studley Park, empty not just of terrestrial mammals but of small birds. It's an over-simplification to say that human degradation of the environment has been universally detrimental: some species have done very well indeed, including some native Australian species; noisy miners and rainbow lorikeets, for instance, proliferate in urban areas, enjoying the ample food provided by street plantings and the open, understorey-less greenery of urban parkland, which suits their requirements and tastes. Brushtail and ringtail possums emerge every night to forage in people's gardens. Bush reserves throughout Melbourne are dotted with nest boxes, waiting for a mammal or bird to come along and occupy them.

Repairing, restoring or recreating an ecosystem is more than just hoping for the best. 'Our approach is to look at entire ecosystems,' the Mt Rothwell website states, 'find the missing links, ensure natural behaviours are preserved ... We are able to adjust our land management practices to the changing environment and climate so that the ecosystem remains balanced.' There's so much work that needs to be done. When I asked my friends how they felt about bringing children into this world, one of my older friends, Jennifer, pointed out: 'Well, we will need someone to wipe our bum and put us to bed when we are old – also neurosurgeons and street sweepers and people able to hoist a jib and ram an earth wall. Also beekeepers and bartenders and tram drivers and tarot card readers and reapers and sowers and gleaners...'

Our guide at Mt Rothwell picks up eye-shine: two glowing points of red. It's a rufous bettong, she tells us, also sometimes called a rat-kangaroo, extinct in the wild in Victoria. As the tour continues we see more and more bettongs – so many that even our guide is surprised. They're wary of us, but not overly so. Our guide tells us that they shelter during the day in a grassy nest on the ground; imagining a bettong sleeping in its flimsy home, I can't help thinking how acutely vulnerable they must be to the sharp senses of foxes and cats. As we continue the tour we see other, similarly vulnerable species, including several eastern barred bandicoots – still surviving in the wild in Tasmania, as is the eastern quoll, but considered extinct on the mainland until a single remnant population was discovered in western

Victoria in the 1980s. With their broad brown and cream stripes, eastern barred bandicoots are easily distinguished from other species when they run in the torch-beam – but they disappear readily into the grass, too.

Bettongs and bandicoots scatter before us as we walk up a gradual hill. At one point our guide pauses by a low outcrop of exposed rock: like much of southern Victoria this is volcanic country, and as our guide knocks on the rock she explains – and we hear – that it's hollow, formed by a bubble of hot volcanic gas. The rocky heights hide another surprise: brush-tailed rock-wallabies, the small southern form of which is vulnerable to fox predation, and as such is critically endangered. Our guide tells us that there are only fifty remaining in the wild.

We continue back down the hill, through the grass and towards the woodland again. The guide gives her torch to my friend Katherine, while she goes to fetch something from a nearby building. Scanning the torch through the undergrowth Katherine catches sight of a different kind of animal, one we haven't seen yet: quick, elusive, dark but covered in white spots. It's gone almost before any of us can see it, vanishing into the understorey of shrubs and grass and fallen logs. When our guide returns she can scarcely believe that we've caught a glimpse of one of the reserve's elusive eastern quolls.

Two years after that visit to Mt Rothwell, brush-tailed rock-wallabies – the NSW form – became one of the iconic animals of post-bushfire recovery efforts: social media feeds filled with photos of the rock-wallabies chewing on carrots and sweet potatoes air-dropped into burned-out national parks, where the vegetation they would ordinarily feed on had been reduced to ash. At the same time, January 2020, the Australian federal government announced $50 million dollars in funding for wildlife recovery efforts. When four months later, in May, they announced another $150 million, it was pointed out that recovery efforts were already being disrupted by the COVID-19 pandemic. Talking to Katherine, who'd spotlighted the quoll at Mt Rothwell, she highlighted the danger that the financial resources needed to deal with the threat of the climate crisis and its attendant disasters such as fire and drought would likely be taken away from the same too-small pool of government funding used to support society's most vulnerable people. It's a hard argument to fault when governments seem hell-bent on hoarding wealth, fixated on the doctrine of surpluses and economic growth.

In March 2017 Don Driscoll, Bek Christensen and Euan Ritchie from the Ecological Society of Australia wrote in *The Conversation* that the Australian federal government had over the previous three years spent $210 million on efforts to save threatened species – compared to $32.4 *billion* on defence in the 2016–17 financial year alone. In June 2020, even as his government was preparing to cut welfare payments in the midst of a pandemic that had made daily life harder than anybody could have prepared for, Prime Minister Scott Morrison announced a further $270 billion in military funding – a thousand times the funding for post-fire environmental recovery – and was quoted as saying that the funding was necessary to prepare Australia for a world that would be 'poorer, more dangerous, and more disorderly'.

Mt Rothwell is privately owned. I can understand the necessity of this model, and I welcome any private individual or company who wants to spend their money on conservation. But I worry too about the neoliberal mindset that is so eager to divest government from any idea of civic, cultural or environmental responsibility; I worry about the generations of voters who've never known any other kind of government: as with ecology, it would seem that politics, too, has a shifting-baseline syndrome.

Like many Australians, I donated to numerous crowd-funding campaigns quickly organised by conservation scientists in the wake of the 2019–20 bushfires, adding what small amounts of money I could to efforts to look for glossy black-cockatoos, Kangaroo Island dunnarts, western ground parrots, spotted-tailed quolls. Like many people I donated also to fundraising efforts for communities particularly badly affected by the fires. Even as people donated, we wondered why the government wasn't spending more money on these or so many other efforts – but we donated anyway. And soon my email inbox began to fill with cautiously hopeful or outright jubilant updates from the campaign organisers, sent to me and an anonymous email list of so many other donors: a surviving population of dunnarts had been found in a tiny patch of unburned habitat; cockatoos had been seen or heard flying over burned forest; a monetary target had been achieved, and there was hope that it could help with rebuilding. As long as there's still something to fight for, we have to fight for it. But government responses were sluggish, and inadequate. And then the pandemic hit.

I don't know how much the human mind is equipped to deal with. In some respects by the time Melbourne's second lockdown was announced several months into the pandemic, I was welcoming the opportunity to retreat into the prescribed world of my home: my bedroom; the sitting-room with its heater; the kitchen; the spare room, where I'd set up a temporary desk to work from home in my day job; and particularly the back garden, where by chance a year earlier I'd built and planted a small vegetable patch. In lockdown, particularly in the anxious early months of the pandemic, home was a sanctuary – a secure enclosure.

First the fires and then the plague had created a whiplash of the unexpected, an emotional fatigue, and though I was forced by government order to stay within these narrow confines – no more than a couple of minutes' walk from one end of the property to the other – I also found myself relishing the smallness of my world, like a snail retreated into its shell. But I knew also that any comfort I felt in this new way of life was built on the suffering of millions of people all over the world, and the deaths of hundreds of thousands and counting. And I found myself frustrated anew by what seemed like the reactiveness and lack of pre-empting in government responses to the virus: how could they not have got ahead of this? How could they not have had a pandemic plan, as my mother, who worked in the Commonwealth Department of Health in the 1990s said they had back then, for the pandemic they all knew was one day going to come? It wasn't till weeks after the states surrounding Victoria closed their borders on us in response to Melbourne's second wave of infections that the true and familiar story started to emerge: of inadequate aged care; of workforce casualisation; of crucial roles privatised and contracted out. COVID-19 has tell-tale symptoms but it is also, itself, a symptom of what continues to ail us. Everything unexpected is bedded within the same old problems.

In the many conversations among scientists I follow on social media, a persistent theme early in 2020 was the difficulty of doing fieldwork during the pandemic. The International Union for Conservation of Nature noted that: 'The COVID-19 pandemic will most likely have significant impacts on wildlife conservation, most of which we will only be able to assess in the longer term. The most immediate of these impacts is the disruption of conservation funding, e.g. from wildlife tourism.' Meanwhile Conservation International stated that:

> There is a misperception that nature is 'getting a break' from humans during the COVID-19 pandemic. Instead, many rural areas in the tropics are facing increased pressure from land grabbing, deforestation, illegal mining and wildlife poaching. People who have lost their employment in cities are returning to their rural homes, further increasing the pressure on natural resources while also increasing the risk of COVID-19 transmission to rural areas. Meanwhile, there are reports of increased deforestation in Asia, Africa and Latin America. Illegal miners and loggers are encroaching on indigenous territories, which could expose remote indigenous communities to the virus. Areas that are economically dependent on tourism face reduced resources as tourism has come to a halt, resulting in a rise in bushmeat (wild meat) consumption in Africa. Meanwhile, illegal mining for gold and precious stones in Latin America and Africa is on the rise, as prices spike and protected areas are left unguarded.

It's easy for those of us in wealthy countries to point the finger at places and people who are less economically secure, but in Australia, seven months deep into the pandemic, the federal government rushed revised environmental legislation through the lower house without debate. In a locked-down world there was more opportunity than ever before to exploit nature while nobody was looking – and not just nature: on 26 October 2020, at the same time as the Victorian premier, Daniel Andrews, was announcing a lifting of lockdown restrictions to a jubilant reception, state government employees were cutting down an ancient tree in western Victoria, sacred to the Indigenous Djab Wurrung people, in order to expand the Western Highway.

There are still conservation works and projects which are ongoing, though – able for one reason or another to keep working, to keep trying to repair the world. One such is on the south coast of NSW, at Booderee National Park at Jervis Bay, 200 kilometres east of Canberra and the same distance south of Sydney. Here, in a coastal landscape, sixty eastern quolls, bred in captivity in Tasmania and in NSW, were released in two stages, in March 2018 and April–May 2019.

One of the partner organisations in this project is WWF-Australia. Rob Brewster was formerly the director of Rewilding Australia, whose mission statement was 'to link Australia's past with Australia's future by restoring our ecosystems', and since that organisation merged with WWF-Australia at the end of 2020 he's been heading up the rewilding unit. His mission is to help 'regenerate Australia' using rewilding strategies following the fires of 2019–20. In an email conversation in 2020 he explains to me why Booderee was chosen:

> Booderee National Park is an approximately 7000 hectare peninsula that is connected to the coastline by a 3 kilometre isthmus. It is a landscape that is not fenced, but rather one that constrains the movement of species by a sea boundary on multiple sides. This provides an ideal landscape to monitor the movement of not only quolls, but species that may interact with, or harm quolls – foxes, dogs, and cats as examples.

Brewster tells me that another, surprising factor that makes Booderee ideal is its lack of cats:

> We certainly know cats have killed and still kill eastern quolls in Tasmania, but in the mesic woodlands of south eastern Australia they are probably not as efficient as they are in arid and rangeland environments. And for whatever reason, perhaps the prevalence of paralysis tick, Booderee National Park just doesn't have cats. They're almost never detected on camera trap in the region, or trapped during the regular park-wide mammal monitoring surveys.

Though, as Brewster points out, cats aren't the only problem in the bush. 'I think foxes have been let off the hook in terms of their devastating effect on small mammals,' he says. 'In most parts of eastern Australia you can mimic the call of a bandicoot in the bush at night and within minutes you'll be able to pick up the eye-shine of a fox (or two) moving in for the kill.'

The fact that Booderee has long been managed for foxes is another reason for choosing it as a place to test the return of eastern quolls.

Which is not to say that it's a quoll paradise, necessarily. Echoing Katherine Tuft's sanguine comments about cats at Arid Recovery, Brewster says: 'The challenge at Booderee is ... accepting that you can't remove every fox from the landscape. And becoming comfortable with the idea that this low level of exposure might actually help our native wildlife learn to persist with foxes over a longer timescale.' Determining exactly where that optimal low level of exposure lies – how many foxes in a landscape are okay – is the big question. And it's a long-term vision: we need to make sure that we're implementing a landscape-management regime that is cost-effective and can basically be continued for the next few decades, year in year out, despite whatever good or bad government we've got in power.

Brewster is cautiously optimistic about the likelihood of eastern quolls living in the wild on mainland Australia in the future – even with the 2019–20 fires and the pandemic. He's philosophical about these challenges: 'The bushfire never reached the park,' he explains, 'and the pandemic has kept our team out of the bush a little. We've just changed our monitoring plans to accommodate. These impediments, however, are probably immaterial to the reintroduction of the eastern quoll when compared to the ever-present threat of the fox.' Brewster tells me that two of the female quolls released in 2018 have now raised young for the third year in a row. However:

> [W]e haven't had adult quolls born in the park that successfully make it through to breeding age, so they have either succumbed to something or have dispersed and we have been unsuccessful in detecting them. Improving monitoring of juveniles as they disperse will be key to determining whether we can sufficiently manage the threats to the survival of young quolls.
>
> We also plan to use what we've learnt to improve survival in the first testing months of living in the wild for captive bred quolls – and this will involve toughening quolls prior to release into the National Park by acclimatising them in a 'semi-wild' fenced environment where they can learn to hunt and scavenge without having to look over their shoulder for introduced predators at the same time.

The quolls aren't the only animals to have been translocated into the park – or the first: Shane Sturgeon and his colleague Gavin McLeod, two of Booderee's Indigenous rangers and members of the Wreck Bay Aboriginal Community Council, which owns and manages the park, tell me that long-nosed potoroos and southern brown bandicoots were introduced into Booderee in 2016 and 2018 respectively. The bandicoots were caught in the wild in state forest near Eden, on the far south coast of NSW. Sturgeon explains that the fact that the bandicoots came from the wild 'is probably a good thing because they'll know how to protect themselves ... [That's] a bit of a worry with quolls being introduced from captivity, whether they [have] the instinct of survival.' Although Brewster had hinted in our email conversation at the scope of the fox-control programme at Booderee, when Sturgeon explains, in detail, the ongoing efforts to control foxes, I'm taken aback:

> We bait [for foxes] every month. We don't stop baiting, we're baiting one part of the park one week – say we'll bait 'em on a Monday – then we'll bait the west, and we'll do that all week, then we'll do vice versa or we'll have a week or two break and then we'll continue every month ... But we've also got cages.

I ask him how long this has been going on for, and he answers matter-of-factly: 'Fifteen years.'

He has some admiration for foxes. 'They're nice animals,' he says sincerely, 'but they're a very devastating animal.' He describes them as 'the biggest headache of all':

> With all the baiting and all the techniques we've been using, it seems like they just come in. They reckon they're not in large numbers, but one fox is still too many ... [You kill a couple] and then

> there's fox sightings next week ... They don't just kill it and eat it like a lion does, it kills and keeps killing like it's a plague.

But McLeod offers some optimism: 'There are quolls and foxes in the landscape [including quolls with pouch young] so they're learning: when they smell a fox or see a fox they go up a tree.'

Sturgeon is keen to find out why quolls disappeared in the first place. He points out that, even before being reintroduced, bandicoots were still present in the park – even after the original quolls were gone. He explains that when he and McLeod were growing up in Wreck Bay they didn't see any of the reintroduced species at all, 'except probably bandicoot'. I ask them if there's community knowledge in Wreck Bay of eastern quolls being present in the landscape – after all, it's within living memory that they were declared extinct on the Australian mainland, and their bones have been found in middens in Booderee. But McLeod is hesitant: he explains that some of the community elders 'are a bit reluctant to pass that knowledge on, so we've got to respect that'.

Sturgeon and McLeod, and their colleagues, worked through the pandemic, even as the park was closed to the public – 'No camping!' they both exclaim down the phone line to me, as if from practice at telling off potential visitors – and McLeod was in the park when the 2019–20 fires burned through NSW: 'We were in the park and watching it come over and the wind helped us out, so [we were] very lucky.' Sturgeon was outside the park at the time, and tells me what he saw: 'The fire was crazy from [Ulladulla] all the way up the east coast.' But after the fires, there were again some signs of hope:

> After the fires we had a big loss of potoroos and southern browns [bandicoots] but in the last two, three months we actually found one potoroo, an old male ... we put out more cameras around the area and in the last month or two we actually found another two potoroos of different size – could be females, which might be good for the old [male] ... Now a couple of potoroos on camera might be a sign of a generation coming through and we actually caught some of the new southern browns in the mammal trapping last season so there's another good indication that things could be happening...

Sturgeon concludes with a note of optimism, the essential component of conservation worldwide: 'We've got our hopes up that there could be new generations coming through.'

Sadly, in March 2021, Brewster gives me some bad news: although the original adult quolls brought into the park were able to live out their full, naturally short lifespans, their offspring had not been so lucky. Domestic dogs, and perhaps a fox, had attacked the juvenile quolls and, Brewster tells me, 'we have unfortunately experienced losses of our perilously small population to the point that we might have lost our final eastern quolls [in Booderee]'.

It's a distressing setback, but setbacks are common in conservation work. I'm reminded of something Brewster had told me in our conversation just a few months earlier:

> We're definitely not at the stage where we can say we've created a self-sustaining population – but I am comfortable with saying that I am more confident than I was a decade ago, that the eastern quoll can be brought back to some of the landscapes it has disappeared from if we continue to test and refine reintroductions into landscapes where foxes and other threats are managed. One thing that I think the reintroduction team has realised is that reintroductions take time! It takes the testing of different strategies, and steps backward that can lead to steps forward. I was buoyed after hearing from Professor Carl Jones from the Durrell Wildlife Conservation Trust, who, at a Taronga Wildlife [Translocation] Health Symposium showed that many of the reintroductions he had led on the islands of Mauritius took an average of ten years to get the population trajectory going the right way.

Or, as Brewster put it more succinctly: 'It's a hard slog trying to reverse the extinction of a species from a landscape.'

4

During the bushfires of 2019–20, inner-city Melbourne felt like the safest place to be. Even the fire that broke out in Plenty Gorge, only half an hour by train from where I live, felt a world away. In the 2020–21 fire season, ads started appearing on TV in Victoria explaining what to do if there was a grass fire near your home: 'Move two streets back.' The urban environment – all that glass and concrete and tarmac – is a buffer against infernos.

But it's no protection from other environmental threats. In Melbourne I see foxes relatively often; the Victorian Department of Agriculture notes that: 'The densities of foxes residing in cities (3–16 per sq km) can be considerably higher than densities observed in farmland in central Victoria (4 per sq km).' The same document explains at length the impact of foxes on Australian wildlife since their introduction in the late 19th century:

> The impact of the red fox combined with habitat degradation is the most likely cause of 'at risk' native animal declines ... Foxes are thought to have played a major part in the demise and extinction of many ground-dwelling native species in the last 130 years. It has been suggested that surplus killing by foxes may have been a major contributor to the rapid mainland extinction or

> reduction of a range of native species in Australia ... as foxes may kill many animals in a night, yet only consume a small amount of each, this will amount to thousands of mammals, reptiles, birds and insects killed each year by a single fox.

And foxes aren't a problem just because of their presence – sometimes their sudden absence, too, can be just as damaging. Writing about fox control efforts for *The Conversation* in 2013, Christopher Johnson from the University of Tasmania pointed out, in relation to the baiting of foxes with the poison 1080, that: 'A drawback is that reduction in fox abundance can result in increased feral cats (which are also susceptible to 1080 but generally do not take baits), because foxes aggressively suppress cats. For some prey species, cats are just as significant a threat as foxes, or more so.' Feral cats have long since replaced 'native cats' on the Australian mainland.

In 2020, before the state borders closed due to the pandemic, and even afterwards when it looked, optimistically, like the closures may only be brief, I had a plan to travel to South Australia. In an email several months after our phone conversation, Katherine Tuft from Arid Recovery explained to me that that reserve makes some of the money necessary for its activities from night-time tours of the reserve, just like Mulligans Flat and Mt Rothwell. I've never been into the true arid zone of Australia – I've never been further inland on this continent than the semi-arid

Mallee of north-western Victoria, a few patches of which are protected in national parks – and I contacted one of my cousins who lives in Adelaide to ask if he'd be interested in joining me for a road trip up to Arid Recovery. He jumped at the chance – but once the borders closed it became a trip in imagination only.

I knew what part of it would look like. I've travelled up the eastern edge of Spencer Gulf, north from Adelaide: seen the flat farming country, the saltmarsh and mudflats, the hills west of Snowtown with their wind turbines. And the Flinders Ranges to the north, huger than anything else around: the plan had been to travel there, too, to put our feet on land that now once again supported western quolls, after more than half a century's absence. And I know what that land at my feet might look like, to some extent. When I was a child my parents had a print in the bathroom, painted by Jan Woodman and published by the Friends of the Botanic Gardens of Adelaide, of the wildflowers of the Flinders Ranges, and when I close my eyes I can still see those flowers, even though they're only paintings: blue *Wahlenbergias* and yellow *Caladenias* and the dramatic red Sturt's desert peas, alongside dozens of others. These flowers help to create an understorey that houses the small animals on which, in the early decades of the 20th century – when my great-grandfather Walter Saddler moved with his family from Hobart to Adelaide and declared himself in a newspaper interview, 'greatly impressed with the possibilities of this State' – western quolls once fed.

When my father was a boy his parents – my grandparents; one of them Walter Saddler's son Bruce – would take him and his two sisters camping in the Flinders Ranges. Later, when he was a young man and newly returned home from London, my father took my mother there, and she tells a story about it: about how it was her first time in Australia, and how she and my father went bushwalking in the Flinders Ranges, and how horrified she was at the spiders between every bush – larger than any spiders she'd ever seen while growing up in England – and how she wondered what she'd got herself into. The Flinders Ranges have always loomed large in my imagination via my parents' youthful memories. (Once while in a London Underground station I saw an enormous poster for South Australian tourism on the wall: a panoramic photo of the bush and mountains, and splashed across it all in giant letters with the confidence that only comes from being utterly, unwittingly wrong: 'FLANDERS RANGES.')

And now the Flinders Ranges are home to quolls again: in April 2014, forty-one western quolls were released in a trial programme in Ikara-Flinders Ranges National Park. The programme is run jointly by the South Australian government and the Foundation for Australia's Most Endangered Species (FAME), a not-for-profit organisation founded in 1993 that funds conservation programmes throughout Australia. After the success of that trial, a year later another thirty-seven quolls were released in Ikara – known in English as Wilpena Pound – a natural amphitheatre in the south of the national park. Another year later even more quolls were released. Even after severe drought the quoll population is thriving: the South Australian Department of Environment and Water reported in April 2019 that fifty quolls had been caught in a week of trapping, twenty-nine of them new individuals. Importantly, the quolls caught were healthy and the population consisted of a mix of both males and females. In May 2020 the Department's Robert Brandle was quoted in a media release from FAME saying that ongoing conservation efforts had allowed the quolls to 'maintain healthy densities in the initial release area and to expand more widely to the edge of their favoured woodland habitats'.

The success so far of the reintroduction has depended on control of feral cats and foxes, with the quolls' young particularly vulnerable to predation. 'Control', of course, is something of a polite euphemism: it means 'killing'. As FAME states on its website:

> The control of feral cats is being undertaken through the use of Felixer grooming traps, Eradicat baits and the involvement of the Sporting Shooters Association of Australia (South Australia branch).
>
> To some this may seem inhumane, yet to do nothing could lead to the extinction of ... the

> Western Quoll in South Australia as well as other susceptible mammals, reptiles and birds.

It's one of the necessary but uncomfortable contradictions of conservation that saving one animal might mean killing another. For people unable to accept this, particularly when it comes to species such as cats, which most of us only ever encounter as beloved pets, conservationists can seem bloodthirsty and heartless. I tend to think of it as a measure of how thoroughly we've made a mess of the natural world that such drastic steps are required as a last resort: the alternative is an Australia left poorer for the loss of its native wildlife. Peter Menkhorst and Frank Knight's *A Field Guide to the Mammals of Australia* lists twenty mammal species as having gone extinct since European colonisation. Such is the ongoing crisis that this number has grown even since the book's most recent edition was published, in 2011. A disproportionate number of these extinct mammal species were found in inland Australia. It sorrows me to know that neither I nor anybody else will ever see a crescent nail-tail wallaby, or a long-tailed hopping mouse or a lesser bilby. Knowing the scale of the loss to date, I would rather cats were killed where necessary so that the Australian landscape can be populated with western quolls and the other species still remaining in their ecological community.

Which is not the same as wishing for the cats, or foxes, to suffer. There's a tendency in Australia to view invasive, destructive species with hatred; it's an alarming and unsettling parallel with the tendency to speak of the need to control Australia's human population, which is often expressed as an environmental need – a question of 'sustainability' – with barely concealed xenophobia and a fortress mentality. I wonder sometimes if the almost fanatical hatred that many Australian people profess for invasive animals – cats, foxes, common mynas, feral pigeons, etc. – is a way of releasing in a socially acceptable way a xenophobia that might otherwise draw some measure of criticism, or force some measure of self-examination.

On the other hand there is the argument, less frequently put forward but out there all the same, that cats and foxes eating native animals is 'just nature' and that any attempt to do something about it is just more human meddling. It's an argument that quite obviously overlooks the fact that the cats and foxes are only eating native Australian animals *because* of human meddling. To put it simply: we made the mess, and it's our responsibility to clean it up.

For a long time the control method of choice on this continent for killing feral or introduced mammal predators has been 1080 poison – and for just as long the use of that poison has been controversial. Supporters of the poison argue that, because it occurs naturally in Australian vegetation, animals which have evolved in Australia are immune to it; critics of 1080 argue in turn that this isn't the case for all native animals, or for domestic dogs that may inadvertently consume poison bait. Exactly how humane 1080 is – how gentle the death that results from it – is also a subject of ongoing debate.

Some of these disagreements seem intractable. Poisoning will always draw visceral and passionate reactions: it's sinister and disturbing in a way that other methods of control aren't, even if those other methods create undue stress. Regardless, poisoning in one way or another is likely to stay as a control method for invasive species in Australia. But one thing that this continent's human inhabitants have never lacked for is innovation. Traditional poison baiting is a blunt-object approach. There is a new way of poisoning cats that promises something much more highly targeted, and therefore less dangerous to other animals. The Felixer grooming trap mentioned by FAME uses a combination of technology and understanding of animal behaviour to poison feral cats. Sensors – two activation sensors and two blocking sensors – detect when a cat, or more specifically a cat-sized and cat-shaped animal, passes in front of the device: if an animal triggers both activation sensors simultaneously but not the blocking sensors, it's identified as a cat based on specific body height, length and clearance measurements. The clearance is a particularly important measurement as native Australian mammals that can otherwise be similar in size to cats – for example, brushtail possums – have shorter legs.

To further maximise the chance of catching cats, the device can also play 'audio lures' – sounds designed to attract cats. When the sensors are triggered by a cat, a poison gel is shot at high speed onto the target animal. Cats, being fastidious groomers, as any cat owner can attest, will notice the gel and lick it off, and in doing so ingest the poison.

Felixers have also been designed to photograph every animal that crosses their paths, which allows the people deploying them to monitor their effectiveness. According to John Read in his book *Among the Pigeons* (Read's company, Ecological Horizons, helped develop the Felixer):

> The first large field trial of Felixers in and around the Arid Recovery Reserve squirted 72 feral cats in 10 weeks and only one 'non-target' kangaroo out of the hundreds of wildlife also photographed by the Felixer. Two of the cats were wearing radio collars and were found dead within hours of being sprayed. Unlikely to thoroughly groom and tolerant of 1080 toxin, the single squirted kangaroo was not at risk. None of the lizards, bilbies, boodies, rabbits or quolls that approached the Felixers were fired upon.

All the same, Felixers are not a complete solution. Katherine Tuft, when talking to ABC North and West SA radio after the trial at Arid Recovery and explaining their efforts to train native animals to avoid cats, pointed out that cats were likely to stay in the

Australian landscape, at least for the foreseeable future. There are simply too many of them, too widely spread – but also, cats and other non-native animals are too ingrained in our lives.

Not all animals that threaten Australian native species are predatory. Horses and cane toads, for instance, are both serious concerns, and ones that might not seem as immediately obvious as foxes and cats. According to a survey conducted in 2019 by the Australian Alps National Parks Cooperative Management Program, Parks Victoria, the NSW National Parks and Wildlife Service, and the ACT Parks and Conservation Service, the number of feral horses – brumbies – in the Australian Alps was 25,318. This was an increase from 9190 horses recorded in the previous survey in 2014 – or an annual increase of 23 per cent.

Like many other animals with a long history of human use and companionship, horses elicit deep emotions from people. In 1858 my ancestor Michael Farrell was wealthy enough to commission a portrait of himself by the Sydney society portraitist Richard Noble – but in fact the painting could be more properly considered a portrait of the horse: in it Farrell is depicted driving a two-wheel Stanhope gig, drawn by a handsome bay horse, and the painting is titled 'The Favourite Gig Horse Napoleon, the Property of Michael Farrell Esq'. Farrell looks notably pleased with himself, and with his horse. I don't know of anyone who isn't stirred by the sight of a horse – and I include myself in that.

When I was in my late teens and early twenties, just past the years of school assemblies and their rote singing of the national anthem, I'd go with my father to the home games of Canberra's new Rugby Union team. Though he grew up in Adelaide – Australian football territory – my father played rugby when he was a young man, and when it comes to sport it's over the two rugby codes, union and league, that he and I have always bonded. This new team had almost immediate success in the tough Super 12 rugby competition (later the Super 14, then 15, then just 'Super Rugby'), which encompassed teams from Australia, New Zealand and South Africa – and Canberra embraced it.

Canberra has always found pride in the success of its sporting teams, seemingly disproportionate to the size of the city, in part because sport gives the city a different identity from the cliched 'three Ps' – politics, porn and pyrotechnics. Canberra's sporting teams have always represented not just the city but also the region: the Southern Tablelands and its numerous towns and communities, all in NSW but closer geographically and psychologically to Canberra than to the state's capital, Sydney. So the pride that Canberrans and people from the Canberra region found in the success of the new rugby team, from its first season in 1997 – also my final year of high school – was reflected in the fact that the team's official nickname was something emblematic of the region: they were called the Brumbies.

When I was a child my family, like many families in Canberra, used to take annual skiing trips to the Snowy Mountains; school ski trips were also a feature of my high-school years and the subject of much excitement. Just as Australian children learn to sing 'Advance Australia Fair', so too do they learn – if not quite in full, then certainly at least the opening lines – Banjo Patterson's poem 'The Man from Snowy River'. It's largely on the back of this poem that brumbies – wild horses, particularly of the alpine regions of NSW and Victoria – have embedded themselves in the national psyche, particularly in the south-east of the continent, where most Australians live.

But horses, recent arrivals to a continent with no native hoofed mammals, are intrinsically alien to Australia. The year after the 2019 survey, and after the bushfires of 2019–20, Professor Jamie Pittock of the Australian National University wrote of flying over the northern part of the alpine Kosciuszko National Park, home to several endangered native species:

> At first, I wondered if the fires may have spared two animals which live in tunnels in the vegetation on the sub-alpine high plains: the alpine she-oak skink and broad-toothed rat [endangered and vulnerable, respectively]. But not only was their understorey habitat burnt, a dozen feral horses were trampling the peat wetlands and eating the first regrowth.

> On the unburnt or partially burnt plains a few ridges over, 100 or more horses were mowing down the surviving vegetation.

Pittock noted the effects of heavy, hard-footed horses on a delicate alpine environment that never evolved to support such animals:

> Over the years, feral horses have carved terraces of trails into the land causing erosion and muddying the stream bank. As more horses congregate on unburnt patches of vegetation after the fires, more eroded sediment will settle on the stream bed and fill the spaces between rocks where the fish shelter. Ash runoff entering the stream may clog the gills of the fish, potentially suffocating them ... We hovered over a key wetland for the [critically endangered] northern corroboree frog that had not been burnt, deep in the alpine forest. A group of feral horses stood in it. They had created muddy wallows, trampled vegetation and worn tracks that will drain the wetland if their numbers are not immediately controlled.

Cane toads, meanwhile, threaten animals in the north of Australia. From an initial release in one tiny part of Queensland, cane toads are now found from NSW across northern Australia all the way to the Kimberley in Western Australia, increasing their range by as much as 60 kilometres every year. They're prolific breeders, able to lay many more thousands of eggs at a time than can native frog species. And, most significantly, adult cane toads are poisonous, and deaths from this poison has led to population declines for native predators – including northern quolls, the smallest of Australia's four quoll species, and the closest geographically to the only two quoll species found outside Australia: Papua New Guinea's bronze quoll and the New Guinean quoll.

In *A Field Guide to the Mammals of Australia* Peter Menkhorst and Frank Knight describe the northern quoll as being the 'size of a large kitten'. It was once found across a large swathe of northern Australia but is now confined to a handful of disjunct populations. The Australian Wildlife Conservancy, a non-government organisation that is involved – through direct ownership or partnership – in the management of thirty different sanctuaries throughout the Australian continent, puts its population at 100,000 and dropping fast thanks to the 'impacts of cats [which] are exacerbated by extensive hot fires and grazing, which reduce ground cover and hence shelter for small mammals'. All that and poisonous toads, too. It's a lot for a small animal to deal with – too much.

But, surprisingly, some populations of northern quolls persist, even in areas where cane toads have become established. Ella Kelly, from the University of Melbourne, tested the theory that these quolls knew to avoid eating the toads, by 'bringing quolls from toad-invaded areas of Queensland (Mareeba and

Cooktown) into captivity and measuring their response to toads'. Kelly tells me, 'I found quolls for toad-invaded areas were far less interested in toads than their naive counterparts from toad-free areas, so we call these quolls "toad-smart"'. This begged the question: could 'toad-smart' behaviour be used to help in the conservation of northern quolls? Kelly set out to explore this possibility:

> I wondered if we could breed toad-smart quolls for release in threatened populations. This would constitute what we call 'targeted gene flow' – whereby you introduce adaptive genes into areas of a species' range that are under threat. For this strategy to work, I needed to determine if toad-smart behaviour has a genetic basis, instead of being passed down through learning.

In order to find this out Kelly needed two test groups of quolls: one group bred from areas that have toads, and one group bred from areas that don't have toads. The results were striking:

> I found that individuals bred from quolls caught from toad-infested areas were significantly less likely to eat cane toads compared to those from a toad-naive lineage. The quolls I tested were all raised in identical environments and had never seen a cane toad – so I knew their aversion was innate.

We're so used to thinking of evolution as something that happens over millennia, particularly in vertebrates, that rapid adaptation like this is startling. Kelly calls it amazing 'but not surprising considering the strong selective pressure acting on the quolls'. Despite the good news, Kelly also points out that most northern quoll populations do go extinct when faced with cane toads. But the fact that some quolls avoid cane toads – and that it's a heritable characteristic – raises new possibilities for northern quoll conservation. In particular targeted gene flow – 'purposefully translocating individuals who carry a favourable trait to areas of a species' range where that trait is useful' as Kelly describes it in her PhD thesis – has the potential to be an exciting new conservation technique. Kelly and her colleagues decided to give it a try:

> We set about testing the idea of targeted gene flow on an island, to see if toad-smart genes were selected for in the presence of toads, and thus help the population survive. In 2017, in collaboration with the Kenbi Traditional Owners and Rangers, we released 54 northern quolls onto Indian Island in the Northern Territory. This island was free of native quolls, and mammalian predators, but had a population of cane toads so we could test selection. The quolls we released were a combination of toad-smart and toad-naive (as well as hybrids), all bred in captivity and trained not to eat cane toads through conditioned taste aversion.

Unfortunately events conspired against the quolls. The list of ordeals the animals suffered, which Kelly catalogues for me, sounds biblical, though it's all too rooted in the world that we've made: fire, cyclone, predation from birds, plus the cane toads whose presence the experiment depended upon. It was too much, and the translocated quoll population collapsed: going from fifty-four individuals in 2017. 'In 2018 we had a population size of approx. 16 individuals, and this year [2019] it was down to four individuals,' Kelly says.

Kelly and her colleagues are continuing their efforts, but even if future translocated northern quoll populations are more fortunate it may not be as simple as just seeding populations of cane-toad-averse quolls throughout northern Australia. Other predators also have to be taken into account, and not just the usual suspects: in 2015 Kelly and other scientists – including Chris Jolly, also of the University of Melbourne – translocated northern quolls from Astell Island, off the coast of East Arnhem Land in the Northern Territory, to Kakadu National Park. Of the twenty-nine quolls that were translocated, twenty-two were trained through conditioned taste aversion to avoid cane toads. The difference in fortunes between those twenty-two and the seven quolls that didn't receive the training was immediately obvious, as the team wrote in a paper on the translocation published in *Austral Ecology* in 2018: 'Toad-naive quolls were almost all killed by toads within 3 days. Toadaverse quolls, on the other hand, not only survived longer but also were recorded mating.' Unfortunately, however, being averse to a particular prey animal didn't prevent the quolls from themselves becoming prey, and the main culprit was not introduced cats or foxes – but dingoes.

It appears that northern quolls can lose survival skills as quickly as they can evolve them: the quolls on Astell Island were themselves descendants of a translocated population, introduced to the island in 2003 in anticipation of the arrival of cane toads on their mainland home. While the quolls in that original 2003 generation, just by virtue of their having survived on the mainland, would have been able to avoid predation, it took no more than thirteen generations for this ability to have been lost in the quolls that were brought from Astell Island back to the mainland in 2015. Perhaps the future for northern quolls lies offshore – much as eastern quolls have survived offshore from mainland Australia, in Tasmania. But as Chris Jolly point out to me, this would be a terrible shame: 'once they're offshore they're tricky to get back.'

It's no secret that it's easier to attract interest in conservation for some animals than for others. People relate more to those animals that are cuter, or to which we feel a stronger connection. When the conservation of one animal is threatened by the presence of another, the relative appeal of each becomes even more significant. Quolls are more appealing than introduced cane toads, but native frogs and rats are likely less appealing to most people than introduced horses.

Horses are charismatic animals, and as with any other domesticated animal any conversation about them can easily become emotive. I'm not saying this as a criticism: I think more conversations about other species should be emotive. For too long, biology as a scientific discipline has tried to maintain a dispassionate and purely objective perspective, and I think this has come at the expense of the deeper cultural and psychological connections that we have with other species; more and more we seek to divorce ourselves from the natural world of which we are both product and part. But it seems to me that in Australia, with its history and also its present of colonialist thought and practice – a colonialism that manifests itself even in the very concept of this continent as a single nation – the emotional connection that many of us feel to wild horses is fraught in a way that it isn't on other continents.

Australians readily load these animals – present here for not even three centuries – with an emotional and cultural significance that we deny to animals that have been here for thousands or even millions of years, animals that are found nowhere else on earth. And, of course, we value the horse because we value its place in the cultural history of European colonisation: which is another way of saying that we value that colonial culture more than we value the unique and ancient cultures of the Australian continent's Indigenous people. You don't need to think about horses to know that this is true, but they're an instructive example. We – by which I mean the Australian cultural mainstream, by which I mean people like me: white, European in heritage – don't make room in our collective national culture for the things that belong here and have been here since ancient times.

There is a danger in reaching too much from thoughts about animals to thoughts about wider societal issues, though. Those of us who feel compelled to write about nature often find ourselves treating the natural world as a metaphor, as something through which to reflect and expand upon our own concerns rather than something which is of itself. An animal can be a lot of things but even so it is, fundamentally, just an animal – which is more than enough. It feeds, it reproduces, it tries to survive, and it's utterly ignorant of the complexities of humanity's global struggles. Yet the animal's plight is, all the same, inextricably connected to those struggles, and, so, connected to our own sense of self. We know this instinctively even though our entire political and economic structure is dedicated to obfuscating it.

I've never seen my local patch of bush reserve in inner Melbourne more busy, more popular, than in the long, slow pandemic lockdown months of mid-2020, despite the fact that Melburnians were limited to only one hour of exercise a day, despite the fact that the lockdown stretched through the coldest, darkest months of the year. In lockdown many people suddenly took up birdwatching, or gardening. If none of this actually materially changed the living conditions of the plants, animals or ecosystems we suddenly found ourselves drawn to, it nonetheless made us more aware of them. An aeroplane flying overhead in 2020, releasing tonnes of carbon into the atmosphere as it went, seemed suddenly like an extraordinary and brutal imposition rather than just part of the world's ambient background noise.

At the start of the pandemic lockdown in March 2020, when, for my day job, I began working from home as offices around the world were shut, a flock of yellow-tailed black-cockatoos were also resident in my part of Melbourne: I'd hear or see them almost daily, flying over my house or calling from the pine trees on the soccer oval where I took my daily walks. Yellow-tailed black-cockatoos are so commonplace in inner Melbourne now that Birdline Victoria, the repository for all unusual bird sightings in the state, no longer accepts reports of them; the fact that they once did indicates how unusual their presence used to be. It's possible that they moved into the city in the wake of the Black Saturday bushfires, and perhaps another wave has similarly arrived after the more recent fires in Bunyip State Park east of Melbourne. Some members of the species, the endangered glossy black-cockatoo, have certainly fled fires: in May 2020 a group of three of them was seen, astonishingly, in the outer south-eastern suburbs of Melbourne, having travelled 400 kilometres from their usual home territory around Mallacoota, the same town evacuated in flames while Scott Morrison urged Australians to raise families.

I went on a first date at Studley Park once in 2019, where the last original eastern quolls in Victoria survived. I was thirty-nine years old, an awkward age to be single, an age where a person's interest in having kids or not as stated on their online dating profile stops being hypothetical and is suddenly blunt, concrete, an instant deal maker or breaker regardless of other qualities. After coffee we went for a short walk along the bush circuit next to the boathouse, talking, getting to know each other in the awkward way of first dates since time began. In the end, it only lasted another few dates.

In her book *Yarra: a diverting history* Kristin Otto notes that Studley Park has historically been a place for men to meet each other for anonymous sex. If you go to Studley Park Boathouse on any sunny day you'll see scores of families or couples enjoying the surrounds: picnicking, walking or just sitting and watching the river flow past. There's a fertile atmosphere in this deep green pocket of inner-city bush. As I think about the families picnicking at Studley Park, or myself going on a date there, every one of us gently but persistently pushing back against the reality of the world, it occurs to me, to my surprise, that there must be people still alive who can remember a time when there were quolls in Studley Park. Perhaps they were a child sitting in the back seat of their parents' car, returning home late at night, and caught a flash of eye-shine in the brightness of the headlights, the glowing of large white spots, of an animal darting across Yarra Boulevard; perhaps they were a young adult on a date in the darkness, far from prying eyes; perhaps they were slightly older again, wandering along the river, worrying idly about the state of the world. 'We thought a nuclear war was on the cards,' a friend a generation older than me tells me, speaking of his own youth. 'We stopped breeding for a decade, then the lucky ones of us became parents in their thirties and forties.' Perhaps my friend's parents worried about polio; before them, perhaps it was the rise of totalitarianism. While all through the generations, numbers of eastern quolls and other small Australian mammals, most of them nocturnal and so hidden from sight, were rapidly dropping.

I wonder to what extent the wildlife of Australia is like the people of Melbourne, in lockdown in 2020: adapting on their own, but also shaped by the pressures they find themselves under. (Some people, of course, and likewise some animals, don't adapt at all, can't find a way to thrive.) I remember a study from 2019 which found that domestic dogs have more-developed facial musculature than grey wolves, which allows them to be more expressive. The study concluded that dogs with more expressive faces –

even today – are more appealing to humans than those with less expressive faces. But whether this feature had evolved in dogs in response to human selection, and whether that selection was conscious or unconscious, remains an open question. It's a fitting analogue for the world we live in now: dynamic, ongoing, but fundamentally uncertain and, for now, unknowable.

My family hold high the values by which I was raised. That goes without saying. When I tell my mum that one of the subjects of this book will be the question of whether to have children or not, when I tell her that many other people are also struggling with this question, she replies that the world of the near future will need people who have been raised to care about the environment, and to care about each other. It's a compelling argument – and one I've made myself – but now it makes me uncomfortable: I'd like to imagine that anyone can come, at any age, to care about things other than themselves; I don't want any children I may have to be like sandbags, piled high against a rising tide of indifference.

Writing in *The Guardian* during the bushfires in December 2019, Gemma Carey related her own experiences of this same sentiment:

> [F]riends would say, 'Don't feel climate guilt. We need a smart, informed generation if we're going to get out of this.' Or, 'It's kids like yours, like mine, that are going to change things.' In these

> comments I see the hope they are reaching for, the way they quiet their mind when they worry 'what world will be left for my child?' as countless women before them have.

And although my mother wouldn't ever phrase it in this way, it's an argument that I've often seen made in harsher terms, too: that all the 'best' people are hesitating about having children, and all the 'worst' people are pushing on with business as usual. What does it mean to divide people into 'best' and 'worst', particularly in this moment of rising global fascism?

> I'm not sure that I want to find out. The world is already too wracked with inequality. As Carey points out in a follow-up to her December article, published a couple of weeks later: In a matter of weeks Canberra ... has shown us how every aspect of climate change disproportionately impacts those who have the least social and economic resources. The city has air inequity, where those with high incomes can purchase thousand-dollar air purifiers, while those who do not sit in their homes marinating in particulate counts that are known to be beyond hazardous. And where people like me can choose not to have surgery in a smoke-filled room, while others are giving birth to babies whose first breath will choke them.

Or as my friend Angela, one of the warmest and most caring people I know, tells me:

> I have always wanted children and for the first time ... I have become more ambivalent about it, precisely for this reason: that by the time they are twenty, thirty ... that existence will be harder for them, and that so much of what exists now may not exist anymore *and* that I may not ever have the kind of income that would help to protect them ... we know the 'haves' will be able to 'adapt' to climate change better than those who are not cushioned with money.

I hear an echo here of the way that it's easier to raise public sympathy for the plight of charismatic, cute animals than for less-appealing subjects.

But I wonder, too, if the emphasis on parents instilling values in their own children is itself part of the problem: if it reinforces an atomised society, in which every family is its own unit, discreet from all others. I remember many years ago my father, apropos of nothing as far as I can recall, reflecting on modern divorce rates and speculating that it may be because parents nowadays are expected to do everything themselves; he pointed out that he and my mother, when my brother and I were young, were able to depend on neighbours or friends to babysit for them when they needed a break.

My friend Sarah has a long-term plan to live in a rural commune, surrounded by friends and their families. As I share rent with a housemate who was a complete stranger before he moved in, and as I look at so many of my contemporaries who are also still share-housing into their forties or older, this more communal – and community-minded – way of living seems more and more realistic. The world now is not one that anyone can survive alone, without help – but it's a symptom of what ails our world that we expect anybody to be able to do so. Even the most ardent survivalists learn what they know from other people. So why should the raising of a child be considered something done only by one or two adults to whom that child is genetically related?

Why, for that matter, should an animal that depends on the intervention of people for its survival not be considered wild? I remember a comment from a birdwatcher in a Facebook group several years ago, about orange-bellied parrots, saying that the species – Australia's most endangered bird species – could no longer be considered wild, such was the level of human intervention in its life; therefore, seeing it in the wild wasn't worth celebrating. It's a stance I disagree with strongly: it seems to encapsulate so much of what's wrong with how we view wildness, and what it means for human and non-human animals to coexist, and our own place in the greater wild.

Northern quolls translocated from place to place, manipulated into breeding with others of their species who have picked up new survival skills, are no less wild, I think, for that manipulation. Likewise, the spotted-tailed quoll that stole the fruitcake from out of Geoff's tent on the South Coast Track in remote

Tasmania was no less wild for having learned how to scavenge from camp sites. The successes of conservation are no less an achievement because they're built on the trial and error of other people's work. Walking for a full week along the South Coast Track, carrying everything we needed on our backs, was no less of an achievement because we walked on a track made or maintained by other people, or because we shared the load of what we needed – food, tent, first aid, etc. – among us. A world cared for in the future will be no less cared for because the children who are taught to care for it are taught by their school teachers, or neighbours, or family friends, rather than by their parents. We understood these things not so long ago; it shouldn't be beyond us to relearn them.

5

When Europeans arrived on the Australian continent and started fundamentally changing the landscapes and ecosystems here, spotted-tailed quolls and eastern quolls overlapped for much of their distribution. Unlike eastern quolls, spotted-tailed quolls – the largest of all quoll species – are still clinging on on the Australian mainland. One place in particular where they can be found is in the northern tablelands of NSW. In an email conversation with me, Guy Ballard, of the University of New England, describes this area as the spotted-tailed quoll's stronghold in the state. When I contact him it's at the end of January 2020, in the very early days of post-bushfire recovery: the fires in northern NSW had been some of the earliest to start in the long, terrible season of 2019–20, burning from September 2019 – well before what we call summer, despite the fact that collectively the fires of 2019–20 are now known as 'Black Summer'. Northern NSW has the highest average rainfall of any part of the state, but by September the bush had been bone-dry for months: when I visited Byron Bay on a family holiday in February 2019, the famous Minyon Falls, 100 metres high, had not a drop of water flowing over.

All the same, the spotted-tailed quolls seem to have coped with the hardships. When I talk to Ballard he and his colleagues are in the middle of monitoring the quolls with remote camera traps – in a fortunate piece of timing, they'd been due to replace their cameras anyway – and so far it's good news: 'Quolls generally seem to have done well after the bushfires,' he tells me:

> [P]articular[ly] where there's rocky refugia. This mirrors observations from researchers at other sites in previous bushfire scenarios. [We're] only missing one female we expected to re-trap and [we] think she's likely moved into a neighbouring patch of forest that didn't burn. Ones we did catch had babies on-board: a good sign.

I ask Ballard why the northern tablelands are so good for the quolls, and the answer is simple: habitat. 'We have lots of contiguous forested areas across public and private land.' Explaining why quolls in particular are coping well in the aftermath of the fires, Ballard is blunt:

> Quolls emerged from burnt areas to a BBQ. Dead animals kept them going in the short term and now they're doing what they do best: hunting and scavenging. Quolls eat every other animal, it just depends on whether they're available at the time. E.g. as macropods continued to die from poor health post fire and drought quolls benefited. As we received rain, and insects started to take off ... great news for quolls again.

A few months later, in May 2020, Ballard gives me an update: 'All of our pre-fire quoll cameras are back out now and we're doing a small amount of trapping too. Some juveniles are being captured (male and female) as well as individuals we haven't seen for 12 months. Happy to see that.' By then the COVID-19 pandemic had started to sweep across Australia, affecting how – and if – everybody worked. I ask Ballard about how he and his colleagues are coping with this:

> We are still monitoring, with careful adherence to safety guidelines. We are balancing the apparently competing demands of 1) social distancing and 2) having two people in the field at a time, by using one car per person, having clearly delineated tasks and not sharing any equipment. We're essentially much safer (to each other and the broader population) than going shopping for loo paper.

Ballard and his team's findings are hopeful news for the spotted-tailed quolls. A map of confirmed sightings of the species on mainland Australia indicates a fragmented distribution, widespread still from Queensland to Victoria but patchy in their known locations – unsurprisingly, given that their habitat, too, is patchy. But that's not to say that spotted-tailed quolls haven't been rumoured to be more widespread. Science has more stringent requirements of evidence than does ordinary day-to-day life, particularly when it comes to rare animals, and though residents of regional Victoria, hundreds of kilometres south of Ballard's study site, are adamant about the presence of quolls around them, scientists – not to mention governments – require something a bit harder.

Which is why there was excitement when, in September 2018, spotted-tailed quoll scats were found in Kinglake National Park, just a 64-kilometre drive from the Melbourne CBD. I once caught the train from my home in Melbourne's inner north out to the last station on the line, Mernda, only forty minutes away; from there I rode my bike half a dozen kilometres further north to Yan Yean Reservoir, Melbourne's oldest water reservoir. I was only an hour from my inner-city home – less if I'd been travelling by car – but at Yan Yean I was almost within touching distance of Kinglake: only 13 kilometres away in a straight line.

We're so used to seeing wildlife on TV, or in photos or videos online, or in guide books or as outlines on a road sign, that we tend to think that if it exists it's *way out there,* somewhere, far away; when I tell people that I regularly go to see wild platypuses in the Yarra River in Melbourne's suburbs, or sugar gliders in a suburban park, they're astonished. And it's right, I think, to be astonished by wildlife, of any kind: every species is a marvel – a marvel of time and the ongoing trial and error of evolution. But the fact that we're so readily astonished at unusual wildlife living among us, in our very midst, surrounded by our houses and our busy lives, is deeply saddening.

We all know that we live in an age of diminishment: that most species are just a shadow of what they once were. If our baselines are forever shifting, generation to generation, we now live in such an age of information that we can all know, in just a few instants, what the old baselines were: what the abundance of life once was. We know that we – the collective 'we' of humanity: our way of life, our dominant social and economic models, the avarice of the greediest and most wealthy of us – are the cause of the downfall of the world around us, even if we more often than not don't want to confront this fact. So a wild animal among us is hard to believe, particularly when that wild animal is rarely seen even in places where it's known with certainty to live.

The scats found in Kinglake National Park in 2018 were the first hard, confirmed and confirmable evidence of spotted-tailed quolls in the area since the 1980s, though locals had said that quolls were around. Additionally, the scats were found in the non-breeding season, raising the possibility – and the hope – that they may be evidence of a colony, rather than of just a single male out roaming for mates, far from his usual home.

In any case, there may soon be spotted-tailed quolls roaming the forests around Kinglake in numbers: in February 2020 the Australian Quoll Conservancy (AQC), a non-profit organisation founded in 2014, submitted a draft proposal to the Victorian government for the reintroduction of spotted-tailed quolls to the Wallaby

Creek catchment in the Kinglake Ranges. In a discussion on Facebook the AQC stated that: 'After 20 months of surveillance with field assistance and facilitation from [the Upper Goulburn Landcare Network] and [the Department of Environment, Land, Water, and Planning], the AQC has established the area is extremely suitable for this first-ever reintroduction.'

Spotted-tailed quolls can turn up in unusual places, at unexpected times. They may appear conspicuous in photos, with their red fur and large white spots, but they're more elusive than you might think. I once saw a video of one in Tasmania, in someone's garden, approaching a birdbath to get a drink. It's in a predator's interests to stay hidden as much as possible, and the quoll in the video wasn't in the open but deep in the garden's bushes. Although I knew it was there it was almost invisible: behind the branches of the bushes its outline was obscured, and its red fur bled into the soil and half-shadows. Most strikingly – if a thing almost unseen can be striking – its spots, so dramatic on the exposed animal, blended into the thousands of pale green dots of the bush's leaves: there were only circles, circles and more circles – some larger and some smaller, some paler and some darker. When the quoll moved through the bush, slowly, carefully, it was as if a light breeze was gently shaking the leaves, nothing more.

In recent years one piece of technology has revolutionised field biology, opening up hidden worlds to us: remote cameras. Triggered by movement and body heat, seeing by both day and by night in the infrared spectrum, these small and unobtrusive cameras can be deployed where animals are known or thought to be and left for days at a time before being checked, recording footage of everything that passes by them onto a memory card. Quiet and still, seemingly passive, the devices don't cause alarm to wildlife. The snow leopard, a near-mythical animal so legendary in its elusiveness that its absence, was the subject of a book by Peter Matthiessen – a book which would prove the foundation of the modern forms of not one but *two* book genres, nature writing and travel writing – is now familiar to people all over the world thanks to remote camera footage of the creature in its Himalayas home. People can sit on their phones or tablets or computers and watch live feeds from cameras peering into the nests of ospreys, or sea-eagles, or peregrine falcons. I have a remote camera that I use to see what the ringtail possums who regularly visit my back garden are up to. Remote cameras are relatively inexpensive and readily available – not just to fulltime scientists but to any nature lover.

Western science, like many other pursuits of knowledge, was once an amateur pursuit – which meant that it was largely a pursuit indulged in only by the wealthy, who had enough leisure time to be able to engage in interests that were time-consuming but with no monetary reward. It's to everybody's gain that science as a professional career has opened up to a wider range of people, but we lose something if we view any human endeavour as the domain of professionals only. 'Amateur' derives from the Latin word for 'lover' – which makes the amateur pursuit of anything literally a labour of love. But 'amateur' is more often used in a derogatory sense these days, so we've come up with a new term for this old idea: 'citizen science'.

One Saturday morning, in the cold months of 2019, I volunteered to help researchers monitor bat boxes in the inner Melbourne suburb of Alphington: I carried the heavy tool bags around while the researchers checked each box and weighed and measured the bats found within them, and put up new boxes. They were the professionals; I was there as an amateur. Though it didn't feel like I was doing a whole lot of science – I was happy just to tag along and learn something about bats, and see some up close – I was that day a citizen scientist. If you've ever helped count birds in your local park, or entered data into an online database about the presence or absence of particular species, or taken a photo of regenerating bush and uploaded it to social media with the requisite hashtag, you've been a citizen scientist. But citizen science doesn't necessarily have to be organised by somebody else, or attached to a larger project.

In May 2018 news broke that a spotted-tailed quoll had been sighted in state forest near Erica, in Gippsland, 130 kilometres east of Melbourne and not far from the popular ski fields of Mt Baw Baw. But it wasn't seen in the flesh: it had been recorded by a remote camera, set by a man named Milan Stupar, described by *The Age* newspaper as a 'lawyer and bushwalking enthusiast', who told the newspaper that he'd 'been trying to capture a quoll on film for more than a year'.

VicForests, the government-owned body that runs logging in Victoria, lists twenty-four logging coupes in the vicinity of Erica. The coupes cover logging activity from 2014 to 2020, though only two are before 2018. There are six coupes listed for 2018, all of them for the months after the quoll was sighted – a sighting that was verified by the Victorian government. The state forests near Erica have an area of 13,486 hectares – 135 square kilometres – but spotted-tailed quolls can be highly mobile, particularly males. It's easy to imagine one making its home in an area earmarked for logging; it's easy, too, to imagine a fragmented and remnant population not being detected in pre-logging surveys.

From coal to timber to pasture, Gippsland is a resource-rich area. Australia often boasts about how it's never had a war on its soil – ignoring the fact that one of the very first acts of James Cook and his crew upon landing in Australia in 1770 was to fire upon the people already here; ignoring also the fact that colonists slaughtered Indigenous people mercilessly and destroyed or attempted to destroy their cultures throughout the continent. Even today some people try to justify that violence by saying that it went both ways – in which case, what do you call that but a war? Even by the horrifying standards of colonial Australia, though, the colonisation of Gippsland was particularly bloody – because the riches to be won were so great. As the Scottish writer Cal Flyn, recounting the actions of her ancestor Angus McMillan, who is either celebrated as the 'discoverer' of Gippsland or reviled as the driving force behind a decade of colonial massacres of Gunaikurnai people, writes:

> In 1839 [Gippsland] would have been lush, widely spaced woodland dotted with freshwater lagoons and punctuated with the curling smoke of the fires of innumerable Aboriginal encampments. Today it is a vast expanse of flat, fertile farmland, well occupied by contented herds of cattle like wildebeest on the Serengeti, black as silhouettes and up to their bellies in flaxen grass heavy with seeds.

And Gippsland is home, too, to some of Victoria's greatest forests.

The native logging industry in Victoria is in a state of flux: in May 2020 the Federal Court found that VicForests had been illegally logging areas of the central highlands that were home to the Leadbeater's possum, the state's critically endangered faunal emblem. This finding came only months after the 2019–20 fires had burned a huge area of Victoria's forest: according to a leaked government document, reported by *The Age* in January 2020, '31 per cent of the state's rainforests ... 24 per cent of wet or damp forests, and 34 per cent of lowland forests.'

The argument always put forward in support of logging, and of other extractive industries, is an economic one: they bring jobs to areas otherwise devoid of them, including Gippsland. It's a hard argument to counter. It's all too easy to think of the people working in these jobs as uncaring or indifferent to the environment, or to the state of the planet. No doubt some of them are – just as some people who are considered broadly progressive may be indifferent to issues of social injustice, or sexism, or other urgent concerns. People are not so simple, and it's unhelpful to be reductive.

Nowadays when I think about communities that are reliant on industries that destroy the environment, I reserve my anger for the decades of government, at all levels and from both nominal sides of politics, who have failed – who have flatout *refused* – to future-proof those communities, to diversify the industries that sustain them and to provide other, more long-term opportunities. It's easier by far to double-down on the same old way of doing business, as it becomes less and less sustainable and self-sufficient, than it is to start creating new ways of living – but what else is a government supposed to be for? Questions of environmental conservation and questions of job security are not separate but are inextricably connected. To borrow an old, old environmental slogan: there are no jobs on a dead planet.

I've passed through Gippsland numerous times – it's a huge area, impossible to avoid if you're travelling at all east from Melbourne by land – but before 2019 I'd never really visited it for any length of time. I was there for a week in October 2019, arriving first in Bairnsdale. In that town, on the side of a bridge across the Mitchell River, I saw painted in huge, red capital letters: 'CLIMATE EMERGENCY'. For months beforehand the news had been full of local governments declaring climate emergency, of activists urging governments to do so. I was surprised – and later embarrassed at my surprise – to see such a bold statement of the issue there, in regional Victoria. It's easy for inner-city residents like myself to imagine that everywhere beyond the major urban centres is inherently conservative and regressive – it's the stereotype that gets pushed on us time and again – but of course the front line in these conversations, these struggles, is out *there,* where extractive industries mean something real and concrete, where they're not just an abstraction.

In Melbourne I live only a couple of kilometres from the city's major grey-headed flying fox colony, where up to fifty thousand individuals of the largest bat species in Australia live along the Yarra River. When I arrived in Bairnsdale, my hosts told me, to my surprise, that the town also had a flying fox colony

– but that for the first time in years the colony had been abandoned as the hungry bats dispersed, looking for food. Despite their reputation for destroying fruit harvests, grey-headed flying foxes feed mostly on the nectar of flowering eucalypts, and so act as crucial pollinators for the trees, and a severe drought meant that the trees weren't flowering: there simply wasn't any blossom for the bats or any other animals to feed on. With this in mind I kept an eye on the eucalypts while I was in Gippsland: during my week there I saw exactly one which was flowering, on a single branch, but other than that there were no buds – not even the debris of past flowering events on the ground.

In Bairnsdale I went to an exhibition by Lisa Roberts at the East Gippsland Art Gallery about the flying foxes. The exhibition was small – only one room. On the wall at the entrance was a statement, headed 'Bearing witness to extinction':

> Australia has the highest extinction rate in the world. In the past 200 years, one in every four world extinctions has occurred here. There are currently over 1800 animal and plant communities at risk of extinction right now. Australia is the only first-world country to make the world top-five hotspots for deforestation. If we are worried about the Amazon, we need look no further than what is happening here – just over the hill.
>
> As forests are destroyed, the national population of grey-headed flying foxes has declined from

> around 20–30 million in 1900 to around 300,000 today, a decline of 99 per cent of that earlier population.

The statement then listed four incidents, from November 2018 to September 2019, of mass deaths of flying foxes from starvation and heat-stress.

If you've ever spent time looking after children, let alone raising them, you'll have experienced what we might call question fatigue: the exhaustion of answering a never-ending list of questions, about why something is the way it is, or how something happens, or where, or what. Each question a child asks only leads to more questions. Science is the same, and requires the same kind of child-like curiosity: each question answered is a door that opens onto another hallway full of doors. A spotted-tailed quoll turns up near Erica. Is it a lone individual, looking for a mate? Or is it part of a larger population? How long has it been there? Where else are there quolls that haven't yet been found? Where else might there be quolls, where their habitat is threatened with imminent destruction? People go to a new place to see what's there; the same is true when the destination is one of knowledge, and information. Each question is an act of creation, in a way – which makes science, however and by whomever it may be practised, a profoundly human endeavour.

'That urge towards creation expresses itself in a million ways,' a friend of mine, Rob, tells me. 'Why write a book? Why write a song? Why make a painting? Why have a kid? If we're facing dire circumstances, aren't all of these things indulgent luxuries? But then again, if we give them up, does that make us less human? Are we sacrificing the things that we should be passing along just in case we can escape this crisis?' He and his partner Aimee, also my friend, recently moved from Melbourne to Traralgon, in Gippsland, only 40 kilometres south of Erica. For her it's a return to the area she grew up in. They're both deeply passionate people – about community, and politics, and the state of the world – and they have an infant daughter. 'I wrestled with [the idea of having a child]', Rob tells me. 'A lot. And I still don't know if I made the right choice. But there is something incredibly human and real about continuing on with the most basic and fundamental aspects of our existence even in the face of dire circumstances. If there is going to be a future, we have to continue to try to create it.'

***

As I write this my dog is silently imploring me with his eyes to take him for a walk. My cat's food bowl is on a shelf behind my right shoulder, out of the dog's reach, and from time to time she comes into the room to see if she can cajole me into giving her more food. I grew up with dogs; I've had cats since I was seventeen. I'm forty-one, and if my dog and cat both live their full lifespans they'll be with me until I'm in my fifties – an age that I find almost impossible to imagine, so different does it seem to where I am now. In the isolation and inwardness of Melbourne's COVID lockdown I came to think of us three – myself, my cat, and my dog – as a little family unit, of sorts: when I sat on the sofa to read in the evening they would both curl up with me, the dog next to me and the cat on my lap, and when I woke up every morning they were both on my bed. It's sentimental but domestic animals breed sentimentality: so deeply ingrained are they in our lives, and in our cultures.

Like many Australian children, I grew up with stories about charismatic foxes, cheeky rabbits, snow in December and sunshine and heat in July; I wonder sometimes if the furtiveness of Australian mammals – their diminished status and often nocturnal habits, the vastness of the landscapes they inhabit – gives them less of a presence in the collective consciousness of mainstream Australia: I wonder what it would take to displace that affection for European animals with an affection instead for quolls, and bandicoots, and sticknest rats. I remember when I was younger the push for the 'Easter bilby', and for a while it seemed to be working – at least in that one, highly specific context – but now the rabbits have crept back onto the shelves, and chocolate bilbies are a novelty at best.

When I try to imagine my life a decade or more from now, increasingly that life is no longer in a city. I'm hardly alone in that regard: the 'tree change' has been a trend for years now, so much so that it's become a cliché, and many of my friends and acquaintances have jumped on board. Already the place where I live, a quiet suburb of inner-northern Melbourne tucked off to the side of the main drag, easy walking distance from Merri Creek and Yarra Bend Park and all the restored bush found there, is somewhere that I cherish precisely because it's not like living in a city: the fact that every night flying foxes fly over my house; the fact that I can walk down to the creek at dusk and see microbats flitting about the treetops, or hear currawongs sing; the fact that every now and then I'll hear a commotion from the local birds and look up to see a falcon, or a goshawk, flying overhead; the fact that a few times a year I'm woken up by the laughing of kookaburras – these are not the typical pleasures of a city, and if I can appreciate them here why not appreciate them somewhere further out, without the accompanying background noise of traffic and the crowds of people?

The pull to the country makes less and less sense with each passing year: nobody in Australia will ever forget the sight of towns and properties throughout the south and east of the continent engulfed in flame and smoke in the 2019–20 fires, or can reasonably expect that the same won't happen again and again. Who would want to put themselves in harm's way like that? But desires aren't rational, and if native animals have to learn to live with cats and foxes in the bush then we might have to learn to live with firestorms.

Perhaps the only way forward, while we wait for the long tail of carbon pollution emitted these days to play out its effects for decades to come, as we continue to fight for present-day carbon emissions to be reduced and cleaner technologies to be supported, is to learn to live a more weightless life: to learn to be less attached.

In January 2020 two Australian writers, Emily Gale and Nova Weetman, created an online auction on Twitter to raise money for fire-fighting efforts in Australia. Like many other writers I took part, offering what I could under the '#AuthorsForFireys' hashtag. The main auction item I offered was a guided birdwatching trip to French Island, east of Melbourne: the group of people who won the trip had been helping with fire-fighting efforts in East Gippsland. As it happened the trip took place just weeks before the sudden arrival of COVID-19 led to the first lockdown in Victoria, and elsewhere in Australia.

The people I took were all older than me, and all much more experienced naturalists than I am, and they lived in Bend of Islands – a place I knew only as a name on a map, 41 kilometres up the Yarra from my house, as the river winds. They gave me a small gift package as thanks for taking them to French Island, and it contained some information about Bend of Islands: I learned that it's classed as an 'Environmental Living Zone', which requires among other things that any resident of the area plant only native plants, apart from a small kitchen garden; erect fences only with a permit; and are forbidden from having dogs or cats.

Normally I catch public transport to French Island, but my guests had cars, so we arranged that I'd meet them in Warrandyte, only a few kilometres from Bend of Islands. After we got back from French Island, as I thought back to that morning when I'd waited at the outer-suburban bus stop in Warrandyte, watching the mist rise off the river through the long bark ribbons of the manna gums, I imagined that I could live somewhere like Bend of Islands: just that little bit further out, but far enough. But the banning of dogs and cats gave me pause.

Could I live without pets? Could I live without that close animal companionship, which in 2020 particularly became a source of such comfort to me? If I move to the country then I think that I might have to accept, at the very least, that my current cat – mercurial and bold and affectionate and exasperating, a huge if constantly alien personality – will be my last ever. Even if confined indoors I'm not sure if I could be comfortable bringing a cat into the bush. But could I then find something similar to that companionship, that comfort that my cat and dog bring, in the wild animals around me?

Shortly after my day job shifted to working from home in March 2020, I set up a makeshift home-office in the spare room of the house I'm renting. Previously my housemate and I used this room as a guest room, when friends or family visited from interstate or needed somewhere to stay temporarily – but in 2020 it quickly became obvious that that wasn't going to be necessary. The room faces out onto the back garden, and a large mulberry tree, which droops its branches into a tangle of fuchsia and correa and rose bushes next to the side fence. As the days got shorter and I found my working days stretching into the early night, I'd get used to seeing a ringtail possum clambering through those branches and bushes. I'd seen it many times before – I knew it lived in or near my garden – but never quite with such intimacy, or regularity. As the months in lockdown passed and the days started to lengthen again, and at dusk I found myself in the sitting room after the working day had finished, instead of at my desk, I'd hear scratching from near the ceiling, and my cat would hear it, too: she'd turn her huge eyes and ears towards the noise and fix all her attention on it. And I'd noticed that on the outside of the wall there was a space in the bricks – corresponding to an air vent high on the inside wall of the sitting room – where the grille had fallen out as the bricks of the old house shifted. I got a ladder and climbed it, carefully, tentatively, and found the ringtail possum staring back at me.

The revered English nature writer Roger Deakin lived, famously, in a rural house that was open to the world outside it: 'A comma [butterfly] lands on my pencil-jar and flexes its wings in the morning sun. Butterflies keep drifting in and out at the open door. They fly right through the study and out into the deeper green of the mulberry tree on the other side of the house.' I've thought about his living arrangement often, in the years since I started reading his work. Could I do something similar? How would that look in an Australian context, where to leave your house open may mean inviting in venomous snakes, spiders, blasting heat, flies?

When I post a short video of my discovery of the possum in my wall to Twitter it prompts hundreds of reactions, including people asking me if I'm going to have it removed or warning me of the damage that it can cause to the house. In truth the only thing I want to do to it is put up a more sheltered nest box so that it can be better protected against the wind and the rain, or the sound and smell of my dog and cat on the other side of the thin mesh separating it from the inside of the house. From my desk in the back room I watch as a red wattlebird gleans spiders from the washing line, and caterpillars from my vegetable garden; as I write this now, I'm sitting at the table on the back deck of my house – I call it 'my' house although it belongs to somebody else but, even if I'm only renting the property, it is my home – shivering slightly in the cold wind on the first day of September, but relishing nonetheless the sunshine, and the gleaming of tiny insects in the air. Just last night in one of the animal-watching groups I follow on Facebook, somebody posted a photo of a brush-tailed phascogale – a threatened species in

Australia, in the same taxonomic family as quolls – in Eltham, only half an hour by suburban train from my home. Powerful owls are found all up and down the Yarra River in Melbourne. I was once walking with a group of friends through a large suburban bush park only 20 kilometres from Melbourne's CBD when one of my friends saw a familiar shape in a tree and asked incredulously, 'Is that a koala?' – it was.

In the depths of lockdown, when exercise for everyone in Melbourne had been restricted to only one hour a day and within a strict radius of 5 kilometres from home, I took solace in Merri Creek: only a ten-minute walk from my house (five if I'm not with my dog), the creek is lined with red gums and manna gums, and yellow and red boxes, black wattles and golden wattles, casuarinas, dianella and hop goodenia and correa and tree violet. Small-leaved clematis tumbles over everything, bright green with wide cream-coloured flowers in the coldest months of the year. And all this is new. Thirty years ago – I've seen photos – Merri Creek was cleared grass monoculture, the trees were willows and poplars. It was a wasteland, before community action brought it slowly back to life. Now there are residents of inner Melbourne, myself included, who've never known it as anything other than bushland.

The Australian landscape will never be what it was. Too many species have been killed, each one significant in its own way to the overall environment. Much of the damage is still ongoing, a bitter legacy of colonial and extractive thinking, which can only see the environment as something in the way – but much of the damage, too, is being repaired, like a wound being sewn shut, the scar still visible but the body functioning again. Unlike in the past, there are now many people and communities working to restore what they can. The future is frightening, but perhaps it can also be quietly hopeful: the Australian landscape of the future will be very different from how it is now, and how it was in the past – but perhaps some of those differences will be good, as well as bad.

Driving across inland south-eastern Australian from Canberra to Adelaide when I was a child all those years ago, nobody in my family could ever have conceived of western quolls being back in that country where once they'd thrived. By the time my father was a young boy in the 1950s quolls had disappeared from South Australia, and in that decade both the northern and southern pig-footed bandicoots likely became extinct entirely, joining the eighteen other species of mammal that had been driven to extinction before them since European arrival in Australia. Now, in both my father's and my lifetimes, western quolls have been returned – and perhaps, if I were to have children, those children would never know the red soil of arid Australia without the imprint of quoll feet.

Living in the present moment can feel like living under a rain of blows – with no hope of them ending any time soon. With the 2019–20 bushfires, then the pandemic, resurgent threats to democracy – whatever else may come next – so many of us are feeling worn down to the nub. In the midst of the pandemic it's understandable that very many people – probably most people – want to return to normal, post-COVID. Apart from the fact that vast numbers of people have lost income, and security, and naturally want to regain what they've lost, *normal* is comforting because of its familiarity, even if the normal world of pre-2020 was one in which huge numbers of people were miserable and exploited. And of course that misery and that exploitation hasn't ended just because the world has changed; the people suffering the most at the hands of the pandemic, and more to the point suffering from the political response to it, are those who were already suffering the most before the pandemic: people who are unemployed or in casual work; the homeless; the poor; the marginalised; people who work in areas not deemed sufficiently 'productive'. In the midst of stage-four lockdown in Melbourne, unable under threat of fine to travel more than five kilometres from my home – a home which because I'm only renting I'm liable in a normal year to lose at any time, as I've lost other homes before – I couldn't help reflecting on the fact that I was, all the same, one of the lucky ones: my day job, which is full-time and salaried, with all the perks that entails such as paid sick leave and paid holiday time, had continued, and other than my dog and my cat I have no dependants, no people other than myself in my immediate orbit who I need to sustain. Though I wasn't able to visit my parents or brother in Canberra for nine months of 2020 – and while in lockdown I didn't know when I might next be able to – I'd all the same lost none of my means. I'm a homebody at the best of times and my life had not materially changed very much, and as such I still had the capacity within me to work, and write, while friends and family around me and in my social media feed struggled to do the same.

Yet to feel yourself as part of a community, or even just to want to feel yourself as such, is to willingly feel sorrow and pain and anger for other people; the same is true, I think, if we're to feel ourselves part of the natural world – but this is a heavy, hard burden to bare, particularly for those people working on the front lines of conservation. 'The scale of the destruction has challenged what is fundamentally an optimistic worldview held by conservationists: that with sufficient time and money, every species threatened by Australia's 250 years of colonial transformation cannot just be saved from extinction, but can flourish once again,' wrote a group of conservation scientists in an article published by *The Conversation* in the aftermath of the 2019–20 bushfires. But grief and anger can be catalysts for action, especially when paralysis and inaction would only further deepen the existing problems:

> So many of us are wondering: have lives spent furthering conservation been wasted? Should we give up on conservation work, when destruction can be wrought on the environment at such unprecedented scales? The answer is, simply, no

> ... as we grieve, we must also remember that decades of conservation work has not been in vain. Some populations and species may indeed have been lost in the recent fires – we shall not know until long after the smoke clears. But the conservation efforts of the past mean fewer species have been lost than would have been the case otherwise.

***

Many people, in the face of the ever-increasing threats we've unleashed upon the planet, look to find safe haven for themselves far away from anyone else: on islands, in bunkers, in compounds. But there are layers of privilege inherent in this mentality – not just because only people of sufficient financial means can afford to wall themselves off from the rest of the world, or buy property on which to live out the increasing dangers of the modern world, but less obviously because jobs and services are concentrated in cities. People who depend upon those jobs and services – to pay the bills or the rent, or just to get through each day in good health or something like it – don't have a realistic option of decamping to somewhere remote and secluded. Safety comes in many forms at an individual level, but at a societal level true safety, I think, comes only through community: not through turning away from the world but in leaning into it, ever deeper, ever more messily and completely. The fantasy of an isolationist is one in which everybody else is a danger. It's a fantasy that recurs again and again in speculative fiction, and one which is not necessarily without a basis in reality – but we're social animals, and if we're to find a way out of this then it needs to be *as a society.* I often think of the late American writer Barry Lopez's simple treatise in his final book, *Horizon:* 'I want everyone here to survive what is coming.'

On the website of the literary journal *Meanjin* in February 2020 Tim Hollo wrote:

> Those in power have always known the importance of myth-making to create an 'us' and a 'them' in order to maintain their power and to direct behaviour. One of the earliest and longest-lasting forms of us and them is the idea that we humans are separate from and superior to the natural world.

Hollo is executive director of the Green Institute, which describes its mission as being to support 'broad communitywide change grounded in the principles of ecology, social justice, democracy, non-violence, sustainability and respect for diversity'. As he went on to write:

> In a world rapidly approaching a precipice, we will need more networks of support, more social cohesion, more layers of redundancy, more cooperation and generosity, more flexibility – all those aspects of society which ecology teaches us

> create resilience but which, in our anti-ecological politics, are unvalued, marginalised and outright erased ... we need to recognise that government and economy are no more than tools that we invented and can reinvent ... continuing to work for climate action by demanding that governments and corporations act within the current system is both destined to fail and underplays our hand ... our strategic goals should reach far beyond such pressure and into cultivating the new democratic systems, norms and institutions we need in order to survive and thrive ... The governments themselves aren't the real problem. If we replace the people in the seats, little will change. The problem lies in a system designed as anti-ecological. What's worse, when we demand action of governments, we are effectively buttressing the power of the system, and abdicating our own power.

What Hollo envisages is neither government action, nor individual action, but *community* action:

> What if we used doorknocking and letterboxing to invite people to get involved in existing groups, and invite them to community meetings to co-design their own local climate-positive, social cohesion projects? What if we supported communities to hold formal and informal citizens assemblies, perhaps connected through those local groups, creating space for Indigenous leadership,

> actively including diverse community members, to discuss what they can do to confront and prepare for the climate crisis?
>
> What if each of those assemblies and gatherings sent representatives to regional assemblies, and shared what they're doing through online clearinghouses, so they could learn from and inspire each other, and so they could consciously envisage their actions as vital pieces of collective action which, together, is cultivating the new ecological democratic alternative?

I'm reminded that the word 'community' describes not just a network of human connections but non-human connections as well: 'An ecological community is a naturally occurring group of native plants, animals and other organisms that are interacting in a unique habitat,' as the Australian Department of Agriculture, Water and the Environment put it. I'm reminded also of the fact that efforts to bring back species such as quolls are efforts to repair as much as possible communities damaged by centuries of colonialism and extractive capitalism.

When I was in Gippsland in October 2019, I was a guest of the Stories of Influence festival, which is held every year – until, like so many other cultural events, the pandemic forced it into hiatus in 2020 – in Lake Tyers Beach, a tiny town just east of the popular holiday destination of Lakes Entrance. Over the course of a weekend, Stories of Influence takes place in a community hall – and sitting in the audience, I heard time and again about the awful colonial history of East Gippsland: the massacres of the Indigenous Gunaikurnai people, which is still an open wound. In 2020 those wounds briefly rose to public prominence in the wake of the Black Lives Matter protests in the United States and here in Australia; in October 2019 at Stories of Influence I was deeply moved to see elderly white people – the group of people who I'd been led to believe by everything I'd absorbed till then would be the most socially regressive – shed tears as they spoke about the urgent need to acknowledge and face up to the role their ancestors played in slaughtering Gunaikurnai people in order to colonise East Gippsland. Colonisation is by design a violent process. But the community of people who make the Stories of Influence festival a success are busy imagining a gentler way of living on – and with – this land. After each day of the festival was finished, I left quietly and walked for a quarter of an hour down the hill from the hall to Lake Tyers – or Bung Yarnda, which I soon learned is its original name in the Gunaikurnai language. I was staying on the lake, as softly as I could, aboard FLOAT.

The easiest way to describe FLOAT is that it's a houseboat – or perhaps house-raft would be a better description: it can't go anywhere on its own without being towed. Moored in a quiet corner of the lake, surrounded by trees, it can only be accessed by boat, or by a punt that connects it by rope to a small jetty 20 metres across the water. As the wind blows across the lake, FLOAT moves with the water, rotating back and forth gently as if keeping watch over the lake. FLOAT is a self-contained, live-in studio, designed and built as an artists' residence and funded to the tune of $350,000 by the Victorian Government's Small Town Transformations project, which 'invited small towns [of fewer than 2000 people] across Victoria to imagine what transformation might mean for their town'.

I was in residence on FLOAT for a week. When I first arrived on board I was shown the ropes by Gary Yelen, a local artist and one of the people responsible for conceiving of and building the vessel. He explained everything to me: the solar power system, the gas cooker, the rainwater tank, the composting toilet. Water from the shower and from the bathroom and kitchen sinks goes directly into the lake, and Yelen pressed upon me the importance of putting as little down the pipes as possible. In my brief time living on FLOAT, listening all day long to the songs of birds in the trees, I thought more constantly and more intently about the direct impact I had on the environment than I have at any other time in my life. Far from being a burden, though, this consciousness was strangely refreshing: there were fewer filters there, both literally and figuratively, and life was more direct.

The locals connected with FLOAT, who went out of their way to make my stay comfortable, told me that the lake was as low as they'd ever seen it. On one of my days on the lake I borrowed one of FLOAT's kayaks and paddled down towards the beach; in wetter years the lake sometimes floods and joins the sea, but now the water was so shallow that behind the beach I became stranded on several occasions on sandbars. I'd heard that, further up the lake, azure kingfishers could be regularly seen – but while on FLOAT I was told that the lake had receded so much in the drought that the branches on which they perched to watch for fish now hung over drying mud rather than water, and they hadn't been seen for a while. But still, the lake supported life: every night on FLOAT I opened the windows and went to sleep to the sounds of fish jumping out of the water – or, more accurately, the sounds of them landing back in the water. One morning I was awoken by an almighty honking and hawing and went onto the deck to see, on shore, a cormorant wrestling a fish out of a pelican's bill, the two birds fighting in a tug-of-war over the catch. The trees surrounding the lake rang with bird calls. And when the sun went down I heard from the trees, well beyond the reach of my torch beam, animal sounds I'd never heard before.

East Gippsland is the stronghold of spotted-tailed quolls in Victoria. On my first day on FLOAT I met another local artist, Josephine Jakobi – Josie, as she was introduced to me. During my stay she told me that a friend of hers in town had seen a quoll in his garden just recently. Josie and I became fast friends:

she's the unofficial custodian of the vessel, occupying it between residencies when it would otherwise be empty, and she helped iron out many of the early kinks in the vessel's design. When I arrived on FLOAT she had several sheets of linen suspended beneath the surface of the lake, 5 or 10 metres behind FLOAT – an art project in which the lake and its ecosystem were imprinting themselves upon the fabric.

Josie grew up by the lake: she's a true local. On my last night on board she took the punt across to FLOAT, bringing a bottle of wine with her, and we sat on the deck and drank it as she told me all about the lake: its wildlife, its history. I'd been invited to the lake to talk about eastern curlews, the subject of my previous book, but it was important to me to not come in as an expert: to absorb more knowledge than I imparted, to be a humble guest. It's a skill that I'm forever trying to learn and relearn: to listen more than I speak, and to not speak out of turn. It's a tricky skill to balance against the life of being a writer, and I'm still figuring it out.

I've recently started learning how to identify the plants around me, particularly the eucalypts, and I cringe now when I remember the confidence with which I misidentified species in the early days of this process when my head was swimming with new knowledge that I didn't fully understand. Despite myself I'm still prone to solitude, to thinking that I can do everything myself; even though it goes against everything I believe, my instincts are still those that have been honed for four decades by a culture that glorifies the individual over the communal, self-knowledge over knowledge of others. We call this a 'can-do spirit'; we also, sometimes, call it the 'pioneering spirit', and it would do us good as a society to take the time to reflect on where that phrase comes from on a colonised continent, and all that it implies.

In the cold months of 2020, well into the pandemic, when we'd all got accustomed to livestreams and Zoom seminars, I tuned into a panel discussing possible ways forward into a frightening future. On the panel was Evelyn Araluen, a poet, researcher and editor, and a descendant of the Bundjalung nation, and when a question from the audience was read out asking her specifically if Indigenous cultures can offer any new ways of knowing and living in the world her mouth tightened visibly before she answered, with great patience, that Indigenous cultures don't offer new knowledge but *old* knowledge, ancient knowledge. The draft strategy for Mulligans Flat in Canberra makes a similar point, acknowledging that Ngunnawal people were caring for country there for millennia before European arrival and foreseeing a time in the near future when their knowledge and culture will be an integral part of caring for country again.

The way most of us live now seems so intractable that we forget how new it is. We destroy the environment and along with it our communities because nobody in power in our lifetimes has ever offered another way to be. But it's seemed apparent to me, too, for a long time, that if there's a way out of this mess it will come not from the top down but from the bottom up: while national governments drag their heels on climate change, on environmental restoration, on social justice, local communities everywhere are figuring out their own ways forward. Each way is a light for those who want to follow; and each light makes the world brighter.

I sometimes think that to live is to worry. I'm reminded of a slogan heard often over social media at the beginning of the Trump presidency: *If you're not worried, you're not paying attention.* I'm a worrier at the best of times. There's too much emphasis placed on individual action to stem the tide of the mounting global environmental catastrophes that face us when what's really needed is systemic change, but all the same I find that I'm coming to value *things* less and less, and not only to not value them but to welcome their loss: wine glasses that I don't use are better given to the ex who used them all the time; excess vegetable seeds from the packet that I bought online to plant in my garden are better shared with my neighbours, if they'll have them, or deposited in the seed exchange that somebody at the other end of my suburb started, rather than hoarded for next year or the year after.

It's a relief I can't quite describe to be *free* of things I don't need, though this is also an attempt to push back against my own hoarding tendencies – tendencies that still fill out my shed, and the space under my bed, and my bedroom cupboards, with old objects and bric-a-brac that I don't need now and couldn't possibly need in the future. Hoarding is a fear of letting go but also, increasingly, in the world that we've created and are living in, it's a fear of contributing to the mess: I don't throw things away because if I do *what imprint will they have upon the world?* Hoarding is a way of avoiding landfill, or perhaps more specifically localising the landfill to your own home and so saving the wider world from it. Hoarding can be an attempt to lighten your load upon the world, by carrying it yourself – though the fatal flaw in that metaphor reveals the hopelessness of this endeavour: because carrying a load doesn't remove that load's weight from pressing upon the earth.

We are passing a heavy load down the generations. The pollution that we spill into the oceans and the skies now will make its effects felt far beyond our lifetimes. It takes my breath away to think about how quickly we've got to where we are: when I was a child, which was not very long ago at all, bottled water was unknown – now a million plastic bottles are purchased every minute. But here comes that worry again: because it's all well and good for someone like me, in a city whose drinking water is clean and safe and delicious, to criticise the consumption of bottled water – but for somebody living in a place where access to clean water is difficult or impossible, water in a bottle could literally be a matter of life and death. But then why plastic, when there are other materials? Why bottled water, instead of safer public water? The questions spin on and on.

It's not fair to expect the people most burdened by the inequalities of our world to also carry the cost of healing it – and not only is it not fair but to do so would only perpetuate that inequality: would only further entrench the very systems that we need so urgently to dismantle. Likewise, conservation efforts that call on the world's poorest nations to curb their excesses while expecting nothing comparable of the world's richest nations will not get us out of this crisis: a colonialist mentality can never resolve the global traumas, human and otherwise, imposed by that same mentality. You cannot give back by continuing to take.

Despite all the seeds I buy and continue to plant, I'm a hopeful gardener more than an expert one: a stick-it-in-the-ground-and-cross-my-fingers type of gardener. If I'd been alive at the time of my ancestor Michael Farrell and had kept chickens, the quolls would definitely have got them. During the pandemic lockdown of 2020, when in a panic people everywhere were trying to grow their own produce, and nurseries were emptied overnight, I read an article about someone who had planted a vegetable garden and three months later was self-sufficient; by that stage my own garden was a year old, and self-sufficiency was nowhere in sight. No matter. I did my weekly grocery shop at my local farmers' market, as I had for years – as I'll continue to do for as long as I need to.

One crisp late-February morning in 2020, only weeks before lockdown but with the coming pandemic only a whisper, I was sitting in the garden of the Abbotsford Convent. It was during the convent's monthly farmers' market – a market that only the following month would be relocated kilometres away as the convent, like so many other public spaces, was closed, as crowds were regulated. None of that yet. And I didn't know yet, either, that eastern quolls had clung on not far from where I was, just on the other side of the river over which the convent's garden looks, until only a human lifetime earlier. I was drinking a coffee from one of the stalls, eating a pastry, sitting on one of the garden benches as I had so many times before, and watching honeybees fly in and out of the last of the blossoms before the weather turned cold and grey. And as I watched them in that perfect moment of peacefulness and contentment I realised that the happiness I feel in any given encounter with wildlife runs parallel with the awareness that such wildlife is a shadow of what it was in my parents' time, or my grandparents' time. The sorrow is still there, but crucially so is the joy. So why should I expect this to be different in any child I raised? Could I not raise them in the joy of tiny moments?

The convent is a place of community, and peace, and respite. It's also a place of wounds and lasting damage. When it was still a convent, and not the mixture of community spaces and restaurants and artists' studios that it is now, it was home to a

Magdalene laundry. In other words, the convent – once a place of confinement for 'women placed in the convent, voluntarily or involuntarily, for being unwanted, neglected, suffering from some disability, destitute, uncontrollable, or for being "wayward" ... [or] orphans, wards of the state, and girls considered by the state or the Church to be in "moral danger"' as the National Heritage List description on the website of the federal Department of Agriculture, Water and the Environment puts it – is a place of trauma. Writing of the convent's laundry in 2014, Audrey K. Hulm, noting that such laundries persisted around the world until the 1990s, said that:

> What is staggering to me is the false moral high ground that is brandished to justify racist Western attitudes to other cultures, whose social systems are painted as far crueller and less refined than our own. Yet here we are as late as 1996, punishing women for sexual activity by way of cruel humiliation and gruelling slave labour.

While the racist attitudes Hulm describes were being imprinted upon the Australian continent the same attitudes were seeing the 'degenerate' marsupial fauna of this continent dwindle further and further.

But the convent, stripped now of its religious associations, is also a place of healing, as all community hubs are. The English nature writer Melissa Harrison says, in the podcast she created to help people struggling through the pandemic in the UK, 'I don't think healing is about erasing what happened. Healing is about learning to live with it, and making it a part of who you are now in a way that feels safe.' The honeybees I was watching in the garden of the Abbotsford Convent were European honeybees, *Apis mellifera,* likely from the hives of the Collingwood Children's Farm, which is next to the convent. European honeybees are one of the oldest domesticated insects in the world – but they can also be a feral animal in many countries, including Australia, where they can occupy increasingly scarce tree hollows upon which native animals depend and also out-compete smaller, more cold-sensitive native bees. I knew all this as I watched them – but I still found pleasure, for a few quiet moments, in the hum of their flight, their industry, their simple *presence.* Harm and healing sit side by side, and I'm not sure if we can talk about one without acknowledging the other.

Is it a sign of my own privilege to be able to sit in that place of trauma and revel in the beauty? Perhaps; probably. That's something I think about constantly. On holiday to Far North Queensland in 2019, I'd marvelled at the vast cane fields, so redolent of tropical Australia as it now is, so iconic, and had grinned from ear to ear at the sight of them because *here I was now, in this place* – even as I was aware the whole time that the industry I was looking at was built on the back of slave labour, so-called blackbirding, which saw people from across the Pacific

Islands kidnapped or tricked into coming to Australia to work in the cane fields. How does a person reconcile such conflicting feelings? More worry – but it's good to worry, because *if you're not worried, you're not paying attention,* and no problem can be fixed without paying attention to it.

Is, then, deciding to not have children – if that's the decision I make – also a sign of my own privilege, as a man who is free to make such a decision without fear of social cost either way, and specifically as a white man in a rich country with all the access to education, freedom and other resources that make such a decision easier, as easy as something so deeply personal and life-changing can ever be? And more particularly still is it a sign of my own privilege as an Anglo-Celtic Australian, and therefore someone whose culture – the dominant culture of a colonising people in a colonised land – is not and has never been under threat on this continent?

Honestly, I don't know. All I can do is keep asking. I think of the image of an eastern quoll at Booderee shimmying up a tree to escape a fox. Not trying to flee along the ground, not hunkering down and trying to hide, not standing and fighting: but another way. Too often we read about proposed high-tech solutions to the current environmental and social crises that we face – which are really just different manifestations of the same crisis – but these 'solutions' are nothing of the sort, because in the end they're just ways of allowing ourselves to continue doing what we're doing, when doing what we're doing is the entire problem to begin with. To get ourselves, and all the other species we share the world with, out of this mess, to make sure the world remains safely habitable for all the future generations of humans and non-humans alike, we need to change what we're doing completely.

I think also, again and again, of FLOAT, and all the ideas and plans bubbling around it: something *collective,* by the community and as a gift to the community. Because I think the point is not for an individual to answer any of the questions that we now face, but to keep asking them so that they can be answered collectively, and to be honest and brave in confronting whatever answers may come – because we have no choice but to find a way forward, and the future is not going to be something that any of us can face alone.

Eventually, in Melbourne in 2020, the lockdown ended, and the borders re-opened, and I was able to go to see my family in Canberra again. On my last day there before returning to Melbourne, my father and I went for another walk in Mulligans Flat. We walked first along the Old Coach Road, which had been the main colonial thoroughfare through the area in the 19th century, past the old woolshed and the various small dams that once would have quenched the thirst of introduced livestock and now are home to turtles. We took a turn down a side-track, through a gate, and found ourselves walking through a field of kangaroo grass as high as our hips, dense and honey-brown and heavy with seed, swaying in the wind like fields of wheat, covering the ground and the nearby hills; native lilies and bluebells and daisies growing in the grass's midst. And everywhere at our feet, on the dirt track through the grassland, shocking and delighting us in its abundance, were the defecations of some small mammal. A carnivorous mammal, because we could see in the scats the colourful jewel-like wing casings of Christmas beetles. I sent a photo to Belinda Wilson to confirm my suspicions.

The air was hot, and quiet, and my father and I were walking through a field of native grass in which the night before, for the first time in decades and leaving plentiful evidence behind, eastern quolls had hunted for food.

Notes

5 foot 6 inches in height: Carter, Max. 2009. "Lunchtime talk 'The favourite gig horse, Napoleon, the property of Michael Farrell Esq.'" Speech, Art Gallery of South Australia, Adelaide. March, 2009.

150 prisoners: The Sydney Gazette and New South Wales Advertiser, 1816. "Sydney; Sitting Magistrate...", December 28, 1816. https://trove.nla.gov.au/newspaper/article/2176973?searchTerm=surry

night owls of every description: The Manaro Mercury, and Cooma and Bombala Advertiser, 1901. "Conference on the Rabbit Question", June 10, 1901. https://trove.nla.gov.au/newspaper/article/119043012?searchTerm=%22tigercat%22

a monster tiger cat: ibid

an unfortunate incident: Mickle, A. 1974. "Lyall, William (1821–1888)." Australian Dictionary of Biography 5, National Centre of Biography, Australian National University. http://adb.anu.edu.au/biography/lyall-william-4051

stocked with kangaroo: Rudé, G. 1967. "Meagher, Thomas Francis (1823–1867)". Australian Dictionary of Biography 2, National Centre of Biography,

Australian National University. http://adb.anu.edu.au/biography/meagher-thomas-francis-2440

one of the commonest: Wheelwright, Horace William, 1861. *Bush wanderings of a naturalist: or, notes on the field sports and fauna of Australia Felix.* London: Routledge, Warne, and Routledge. https://doi.org/10.5962/bhl.title.33268

very destructive: ibid

The Domestic Cat sometimes: ibid

was once found throughout: Australia. National Parks and Wildlife Service (NSW). 2003. The Bioregions of New South Wales – their biodiversity, conservation and history. Chapter 8. https://www.environment.nsw.gov.au/resources/nature/BioregionsContentsIntro.pdf

relatively abundant: Australia. Department of the Environment and Heritage, 2004. Quolls of Australia. https://www.environment.gov.au/system/files/resources/8ad3915e-57ee-4eb8-afd0-4f6bed036896/files/quolls.pdf

31*there is an isolated colony:* Menkhorst, P (ed) 1995, *Mammals of Victoria: distribution, ecology and conservation,* Oxford University Press, Melbourne

one of the commonest of all the bush animals: Seebeck, John. 2004. "Eastern Quoll." *eMelbourne,*

School of Historical & Philosophical Studies, University of Melbourne, accessed 29 March 2021. https://www.emelbourne.net.au/biogs/EM00500b.htm

land which is wild: Shorter Oxford English Dictionary. 2007. 6th edition, 2. Oxford University Press.

unlike any fire Queensland: Doman, Mark. 2018. "From space, the ferocity of Queensland's bushfires is revealed." *ABC News,* December 8, 2018. www.abc.net.au/news/2018-12-08/from-space,-the-ferocity-of-queenslands-bushfires-is-revealed/10594662?nw=0

You have to wonder: Lloyd, Shelley, 2019. "Gold Coast hinterland fire prompts question—can rainforests burn?" *ABC News,* September 9, 2019. https://www.abc.net.au/news/2019-09-09/gold-coast-hinterland-bushfire-why-rainforests-burn/11491364

birds, reptiles and mammals: Australia. Department of the Environment and Heritage, 2004. *Quolls of Australia.*https://www.environment.gov.au/system/files/resources/8ad3915e-57ee-4eb8-afd0-4f6bed036896/files/quolls.pdf

300,000 animals are killed: Jones, Ann. 2015. "Welcome to Tasmania, the roadkill capital of the world." *ABC News,* December 14, 2015. https://www.environment.gov.au/system/files/resources/8ad3915e-57ee-4eb8-afd0-4f6bed036896/files/quolls.pdf

The carnivores: ibid

It is extremely: Bowman, David. 2016. "Fires in Tasmania's ancient forests are a warning for all of us." *The Conversation,* January 29, 2016. https://theconversation.com/fires-in-tasmanias-ancient-forests-are-a-warning-for-all-of-us-53806

There are two stories: Australian Fire and Emergency Service Authorities Council. 2019. *AFAC independent operational review: a review of the management of the Tasmanian fires of December 2018* – March 2019. www.fire.tas.gov.au/userfiles/AFAC/AFAC_Review.pdf

The summer of 2018–19: Australian Fire and Emergency Ser vice Authorities Council. 2019. *AFAC independent operational review: a review of the management of the Tasmanian fires of December 2018 – March 2019.*www.fire.tas.gov.au/userfiles/AFAC/AFAC_Review.pdf

Warrens that had housed: Read, John. 2019. *Among the Pigeons: Why our cats belong indoors.* Adelaide, South Australia: Wakefield Press, South Australia.

The Arid Recovery: ibid

Subsequent expansions: ibid

to be 'agile': The Sydney Morning Herald. 2015. "Malcolm Turnbull and Julie Bishop address the media following leadership spill." September 15, 2015. http s://www.smh.com.au/politics/federal/malcolm-turnbull-and-julie-bishop-address-the-media-following-leadership-spill-20150915-gjmjyk.html#ixzz3llSlEUGv

how to exist: Carvan, Tabitha. "How do we go on?" Accessed April 27, 2021. https://health.anu.edu.au/news-events/news/how-do-we-go

We know enough: Crist, Meehan, 2020. "Is it okay to have a child?" *London Review of Books.* February 23, 2020. https://www.lrb.co.uk/the-paper/v42/n05/meehan-crist/is-it-ok-to-have-a-child

The problem we're facing: Carvan, Tabitha. "How do we go on?" Accessed April 27, 2021. https://health.anu.edu.au/news-events/news/how-do-we-go

[S]hift[ing] responsibility: Crist, Meehan, 2020. "Is it okay to have a child?" *London Review of Books.* February 23, 2020. https://www.lrb.co.uk/the-paper/v42/n05/meehan-crist/is-itok-to-have-a-child

[He] was three: Carvan, Tabitha. "How do we go on?" Accessed April 27, 2021. https://health.anu.edu.au/news-events/news/how-do-we-go

Some people say: Crist, Meehan, 2020. "Is it okay to have a child?" *London Review of Books.* February 23, 2020. https://www.lrb.co.uk/the-paper/v42/n05/meehan-crist/is-it-ok-to-have-a-child

Recently, the logic: Crist, Meehan, 2020. "Is it okay to have a child?" London Review of Books. February 23, 2020. https://www.lrb.co.uk/the-paper/v42/n05/meehan-crist/is-it-ok-to-have-a-child

We live in capitalism: Le Guin, Ursula, 2014. Ursula's acceptance speech: Medal for Distinguished Contribution to American Letters. National Book Awards, New York City. November 20, 2014. https://www.ursulakleguin.com/nbf-medal

Any human power: ibid

Last year more: Mulligans Flat Woodland Sanctuary, 2016. Sign-up for Turtle Patrol this Spring." August 8, 2016. https://mulligansflat.org.au/sign-up-for-turtle-patrol-this-spring/

rebuilding natural: Mulligans Flat Woodland Sanctuary, 2020. *Mulligans Flat Woodland Sanctuary Draft Strategy 2020–2045 – for public comment.* ACT Government. https://www.ursulakleguin.com/nbf-medal

reintroduced native fauna: ibid

golf-ball sized: Bladen, Lucy. 2020. "Canberra hail storm: Wild weather lashes ACT causing wide-ranging damage." Canberra Times, January 20, 2020. https://www.canberratimes.com.au/story/6588613/heavy-storms-and-huge-hailstones-lash-canberra/

Even though Yellowstone: Farquhar, Brodie. 2020. "Wolf reintroduction changes ecosystem in Yellowstone." Yellowstone National Park, June 30, 2020. https://www.yellowstonepark.com/things-to-do/wolf-reintroduction-changes-ecosystem

have multiple effects: ibid

While ignorance obviously: Dyett, Jordan and Thomas, Cassidy. 2019. "Overpopulation Discourse: Patriarchy, racism, and the specter of ecofascism." *Perspectives on Global Development and Technology* 18: 1–2, January 18, 2019. https://doi.org/10.1163/15691497-12341514

In March 2017: Driscoll, Don, Bek Christensen and Euan Ritchie, 2017. "Government needs to front up billions, not millions, to save Australia's threatened species." *The Conversation,* March 21, 2017. https://theconversation.com/government-needs-to-front-up-billions-not-millions-to-save-australias-threatened-species-74250

A drawback is that: Johnson, Christopher, 2013. "Is it too late to bring the red fox under control?" *The Conversation,* February 13, 2013. https://theconversation.com/is-it-too-late-to-bring-the-red-fox-under-control-11299

The first large field trial: Read, John, 2019. *Among the Pigeons: Why our cats belong indoors.* Adelaide, South Australia: Wakefield Press, South Australia.

At first, I wondered: Pittock, Jamie, 2020. "Fire almost wiped out rare species in the Australian Alps. Feral horses are finishing the job." *The Conversation.* January 28, 2020. https://theconversation.com/fire-almost-wiped-out-rare-species-in-the-australian-alps-feral-horses-are-finishing-the-job-130584

Over the years: ibid

size of a large kitten: Menkhorst, Peter and Frank Knight. 2010. *Field Guide to the Mammals of Australia* (Third Edition). Oxford: Oxford University Press.

Toad-naive quolls: Jolly, Christopher, Ella Kelly, Graeme R. Gillespie, Ben Phillips and Jonathon K. Webb, 2017. "Out of the frying pan: reintroduction of toad-smart northern quolls to southern Kakadu National Park." *Austral Ecology,* October 24, 2017. https://doi.org/10.1111/aec.12551

[F]riends would: Carey, Gemma, 2019. "If I have no hope for the planet, why am I so determined to have this baby?" *The Guardian,* December 26, 2019. https://theguardian.com/commentisfree/2019/dec/26/if-i-have-no-hope-for-the-planet-why-am-i-so-determined-to-have-this-baby

In a matter of weeks: Carey, Gemma, 2020. "Being pregnant in a climate emergency was an existential challenge. Miscarriage has brought a new grief." *The Guardian,* January 10, 2020. https://www.theguardian.com/commentisfree/2020/jan/10/being-pregnant-in-a-climate-emergency-was-an-existential-challenge-miscarriage-has-brought-a-new-grief

In 1839 [Gippsland]: Flyn, Carey, 2016. *Thicker than Water.* Pg 113. Sydney: Harper Collins.

A comma [butterfly] lands: Deakin, Roger, 2008. *Wildwood: A journey through trees.* Pg 28. United Kingdom: Penguin.

The scale of the destruction: Garnett, Stephen, Brendan Wintle, David Lindenmayer, John Waoinarski, Martine Maron and Sarah Legge, 2020. "Conservation scientists are grieing after the bushfires – but we must not give up." *The Conversation,* January 21, 2020. https://theconversation.com/conservation-scientists-are-grieving-after-the-bushfires-but-we-must-not-give-up-130195

I want everyone: Lopez, Barry, 2020. *Horizon.* London: Vintage.

Those in power: Hollo, Tim, 2020. "The end of the world as we know it." *Meanjin,* February 3, 2020. https://meanjin.com.au/blog/the-end-of-the-world-as-we-know-it/

In a world rapidly: ibid

What is staggering: Hulm, Audrey K., 2014. "Magdalen laundries for "sluts" in "moral danger": why I founded 'Found Festival'." *Lipmag,* April 23, 2014. https://lipmag.com/culture/magdalen-laundries-for-sluts-in-moral-danger-why-i-founded-found-festival/

I don't think healing: State Bar of Texas. *State Bar of Texas Board of Director's Meeting,* July 27, 2020. http://www.texasbarcle.com/CLE/BDPLAYER5.ASP?sMeetingID=072720

Front Cover Flap

When Harry Saddler first encountered a quoll while camping with his father, he was struck by the beauty of the rare creature who had emerged from the bush, sniffing for dinner. As Harry frantically snapped a photo, the fast-moving mammal disappeared back into the undergrowth.

Many years later that blurry photo remains a memory, as fleeting as the animal it captured. After two centuries of habitat destruction, quolls are now on the brink of extinction and Harry, contemplating fatherhood, aches for the absence of all the species lost to children born today.

Questions Raised by Quolls is an eloquent examination of extinction and conservation set against the backdrop of global climate change. From his own family lineage, Harry reveals how humanity's ever-accelerating modern way of life runs parallel with the destruction of the natural world. Evocative and challenging, this eulogy to lost species will force you to question your place in the vast interconnected web of life.

Back Cover Flap

HARRY SADDLER is a Melbourne-based writer. His writing about the interactions between people, animals and the environment has been published in *The Lifted Brow, Meanjin* and *The Guardian,* among others. His book *The Eastern Curlew* was shortlisted for the Queensland Literary Awards in 2019.

Back Cover Material

'They came at dusk, drawn to our camp site by the smell of food. We were cleaning up after dinner in the gathering dark, the pots and pans scraped empty but with traces of our meal still lingering, when we heard the noise of them: yapping calls, one to the other, or just to themselves in their excitement.'

Printed in Dunstable, United Kingdom

91338348R00118